T0292568

Springer Undergraduate Texts
in Mathematics and Technology

More information about this series at http://www.springer.com/series/7438

Ching-Shan Chou • Avner Friedman

Introduction to Mathematical Biology

Modeling, Analysis, and Simulations

 Springer

Ching-Shan Chou
Ohio State University
Columbus, Ohio, USA

Avner Friedman
Ohio State University
Columbus, Ohio, USA

ISSN 1867-5506   ISSN 1867-5514   (electronic)
Springer Undergraduate Texts in Mathematics and Technology
ISBN 978-3-319-29636-4   ISBN 978-3-319-29638-8   (eBook)
DOI 10.1007/978-3-319-29638-8

Library of Congress Control Number: 2016931866

Mathematics Subject Classification (2010): 92B05

Printed on acid-free paper

This Springer imprint is published by Springer Nature
The registered company is Springer International Publishing AG Switzerland

Contents

Chapter 1
Introduction

The progress in the biological sciences over the last several decades has been revolutionary, and it is reasonable to expect that this pace of progress, facilitated by huge advances in technology, will continue in the following decades. Mathematics has historically contributed to, as well as benefited from, progress in the natural sciences, and it can play the same role in the biological sciences. For this reason we believe that it is important to introduce students very early, already at the freshman or sophomore level, with just basic knowledge in Calculus, to the interdisciplinary field of mathematical biology. A typical case study in mathematical biology consists of several steps. The initial step is a description of a biological process which gives rise to several biological questions where mathematics could be helpful in providing answers. The second step is to develop a mathematical model that represents the relevant biological process. The next step is to use mathematical theories and computational methods in order to derive mathematical predictions from the model. The final step is to check that the mathematical predictions provide answers to the biological question. One can then further explore related biological questions by using the mathematical model.

This book is based on a one semester course that we have been teaching for several years. We chose two sets of case studies. The first set includes chemostat models, predator–prey interaction, competition among species, the spread of infectious diseases, and oscillations arising from bifurcations. In developing these topics we also introduced the students to the basic theory of ordinary differential equations, and taught them how to work and program with MATLAB without any prior programming experience. The students also learned how to use codes to test biological hypotheses.

The second set of case studies were cases adapted from recent and current research papers to the level of the students. We selected topics that are of great public health interest. These include the risk of atherosclerosis associated with high

© Springer International Publishing Switzerland 2016
C.-S. Chou, A. Friedman, *Introduction to Mathematical Biology*,
Springer Undergraduate Texts in Mathematics and Technology,
DOI 10.1007/978-3-319-29638-8_1

cholesterol level, cancer and immune interactions, cancer therapy, and tuberculosis. Throughout these case studies the student will experience how mathematical models and their numerical simulations can provide explanations that may actually guide biological and biomedical research. Toward this goal we have also included in our course "projects" for the students. We divided the students into small groups, and each group was assigned a research paper which they were to present to the entire class at the end of the course.

Another special feature of this book is that in addition to teach students how to use MATLAB to solve differential equations, we also introduce some very basic numerical methods to familiarize the students with some numerical techniques. That will greatly help their understanding in using different MATLAB functions, and can further help them when they try to use other computer languages in the future. Overall, our book is quite different from traditional mathematical biology textbooks in many aspects.

We believe that the book will help demonstrate to undergraduate students, even those with little mathematical background and no biological background, that mathematics can be a powerful tool in furthering biological understanding, and that there are both challenge and excitement in the interface between mathematics and biology.

This book is the undergraduate companion to the more advanced book "Mathematical Modeling of Biological Process" by A. Friedman and C.-Y. Kao (Springer, 2014), and there is some overlap with Chapters 1, 4–6 of that book. We would like to thank Chiu-Yen Kao who taught the very first version of this undergraduate course.

The MATLAB codes (in M-files) for the sample codes printed on the book are available in the Supplementary Material. The supplementary material can be downloaded from http://link.springer.com/book/10.1007/978-3-319-29636-4.

Chapter 2
Bacterial Growth in Chemostat

2.1 What Is a Chemostat

A chemostat, or bioreactor, is a continuous stirred-tank reactor (CSTR) used for continuous production of microbial biomass. It consists of a fresh water and nutrient reservoir connected to a growth chamber (or reactor), with microorganism. The mixture of fresh water and nutrient is pumped continuously from the reservoir to the reactor chamber, providing feed to the microorganism, and the mixture of culture and fluid in the growth chamber is continuously pumped out and collected. The medium culture is continuously stirred. Stirring ensures that the contents of the chamber is well mixed so that the culture production is uniform and steady. If the steering speed is too high, it would damage the cells in culture, but if it is too low it could prevent the reactor from reaching a steady state operation. Figure 2.1 is a conceptual diagram of a chemostat.

Chemostats are used to grow, harvest, and maintain desired cells in a controlled manner. The cells grow and replicate in the presence of suitable environment with medium supplying the essential nutrient growth. Cells grown in this manner are collected and used for many different applications.

These applications include:

1. **Pharmaceutical**: for example in analyzing how bacteria respond to different antibiotics, or in production of insulin (by the bacteria) for diabetics.
2. **Food industry**: for production of fermented food such as cheese.
3. **Manufacturing**: for fermenting sugar to produce ethanol.

A question which arises in operating the chemostat is how to adjust the effluent rate, that is, the rate of pumping out the mixture. In order to operate the chemostat

---

Electronic supplementary material The online version of this chapter (doi: 10.1007/ 978-3-319-29638-8_2) contains supplementary material, which is available to authorized users.

© Springer International Publishing Switzerland 2016
C.-S. Chou, A. Friedman, *Introduction to Mathematical Biology*,
Springer Undergraduate Texts in Mathematics and Technology,
DOI 10.1007/978-3-319-29638-8_2

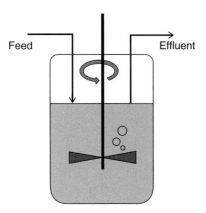

Fig. 2.1: Stirred bioreactor operated as a chemostat, with a continuous inflow (the feed) and outflow (the effluent). The rate of medium flow is controlled to keep the culture volume constant.

efficiently, the effluent rate should not be too small. But if this rate is too large, then the bacteria in the growth chamber may wash out. In order to determine the optimal rate of pumping out the mixture we need to use mathematics. In this chapter, we develop a simple mathematical model in order to determine the optimal effluent rate. A more comprehensive model will be developed in Chapter 8.

The mathematical model will be described by a differential equation. In this book we shall encounter many differential equations that model biological processes. We therefore review here some of the basic theory of differential equations.

2.2 Differential Equations

Differential equations of the **first order** have the form

$$\frac{dx}{dt} = f(x,t), \tag{2.1}$$

where $f(x,t)$ is a given function. Solving this equation means that we have to find a function $x(t)$ which satisfies

$$\frac{dx(t)}{dt} = f(x(t),t).$$

There are in fact many such solutions; but the solution will be unique if we prescribe a condition such as

$$x(t_0) = x_0 \tag{2.2}$$

for some values x_0 and t_0. The system (2.1)–(2.2) is called an **initial value problem**; the initial time t_0 may be taken, for instance, to be $t_0 = 0$.

Example 2.1. The solution of the differential equation

$$\frac{dx}{dt} = t^2$$

is $x(t) = \frac{t^3}{3} + C$, where C is an arbitrary constant. If we prescribe initial condition $x(0) = 5$, then $C = 5$ and the unique solution is $x(t) = \frac{t^3}{3} + 5$.

Example 2.2. The solution of the initial value problem

$$\frac{dx}{dt} = x + 2, \quad x(0) = 3$$

is given by

$$x(t) = 5e^t - 2.$$

There are several classes of differential equations that can be solved explicitly, and they are introduced in the following subsections.

2.2.1 Linear Equations

Linear differential equations have the following form:

$$\frac{dx}{dt} + p(t)x = g(t), \tag{2.3}$$

where $p(t)$ and $g(t)$ are given functions of t. In order to solve such an equation we introduce the integral of $p(t)$,

$$P(t) = \int_0^t p(s)ds,$$

and multiply Eq. (2.3) by $e^{P(t)}$,

$$e^{P(t)}\frac{dx}{dt} + e^{P(t)}p(t)x(t) = e^{P(t)}g(t).$$

Note that

$$\frac{d}{dt}[e^{P(t)}x(t)] = e^{P(t)}\frac{dx}{dt} + e^{P(t)}p(t)x(t)$$

by the definition of $P(t)$. Therefore,

$$\frac{d}{dt}[e^{P(t)}x(t)] = e^{P(t)}g(t),$$

and by integrating both sides from 0 to t, we get

$$e^{P(t)}x(t) - x(0) = \int_0^t e^{P(s)}g(s)ds.$$

It follows that

$$x(t) = e^{-P(t)}x(0) + e^{-P(t)} \int_0^t e^{P(s)}g(s)ds \qquad (2.4)$$

is the solution of (2.3) with prescribed $x(0)$.

Note that in Example 2.2 above $p(t) = -1$, $P(t) = -t$, $g(t) = 2$,

$$e^{-P(t)} \int_0^t e^{P(s)}g(s)ds = 2e^t \int_0^t e^{-s}ds = 2e^t(1 - e^{-t}),$$

and formula (2.4) yields

$$x(t) = e^t x(0) + 2(e^t - 1) = 5e^t - 2 \quad \text{if } x(0) = 3.$$

2.2.2 Separation of Variables

Differential equations with separable variables are of the form

$$\frac{dx}{dt} = g(x)h(t). \qquad (2.5)$$

Rewriting this equation in the form

$$\frac{1}{g(x)}\frac{dx}{dt} = h(t)$$

we get, by integration with respect to t,

$$\int \frac{dx}{g(x)} = \int h(t)dt,$$

from which we obtain the solution

$$G(x) = H(t) + C$$

where

$$G(x) = \int \frac{dx}{g(x)}, \quad H(t) = \int h(t)dt.$$

Example 2.3. Consider the equation

$$\frac{dx}{dt} = \frac{t}{x^2}.$$

Writing it in the form

$$x^2 \frac{dx}{dt} = t$$

we get, by integration,

$$\frac{x^3}{3} = \frac{t^2}{2} + C.$$

2.2.3 Homogeneous Equations

Differential equations that can be written in the form

$$\frac{dx}{dt} = g(\frac{x}{t}) \tag{2.6}$$

are called homogeneous equations. Such equations can be solved by introducing a new variable $v = \frac{x}{t}$, or $v(t) = \frac{x(t)}{t}$. Then

$$\frac{dx}{dt} = \frac{d}{dt}(tv) = v + t\frac{dv}{dt},$$

and Eq. (2.6) becomes

$$t\frac{dv}{dt} + v = g(v),$$

which is an equation with separable variables, namely,

$$\frac{dv}{dt} = \frac{g(v) - v}{t}.$$

Hence

$$\int \frac{dv}{g(v) - v} = \ln t + C.$$

If we denote the integral of $1/(g(v) - v)$ by $K(v)$, then the solution of Eq. (2.6) is given implicitly by the formula

$$K(\frac{x}{t}) = \ln t + C.$$

Example 2.4. The equation

$$\frac{dx}{dt} = \frac{x^2 + t^2}{xt}$$

can be written in the form

$$\frac{dx}{dt} = g(\frac{x}{t}), \text{ where } g(v) = v + \frac{1}{v}.$$

Setting $v = \frac{x}{t}$, we get

$$\int \frac{dv}{g(v) - v} = \ln t + C,$$

and the integral of the left-hand side is $\int v dv = \frac{v^2}{2}$. Hence

$$\frac{1}{2}(\frac{x}{t})^2 = \ln t + C,$$

or

$$x^2 = 2t^2(\ln t + C).$$

2.2.4 Exact Equations

Consider a differential equation of the form

$$g(x,t)\frac{dx}{dt} + h(x,t) = 0. \tag{2.7}$$

If there is a function $F(x,t)$ such that

$$\frac{\partial F}{\partial x} = g, \quad \frac{\partial F}{\partial t} = h, \tag{2.8}$$

then Eq. (2.7) is called an **exact equation**, and it can be written in the form

$$\frac{\partial F}{\partial x}\frac{dx}{dt} + \frac{\partial F}{\partial t} = 0$$

or

$$\frac{dF(x(t),t)}{dt} = 0.$$

Hence,

$$F(x(t),t) = constant,$$

and the solution of Eq. (2.7) is given implicitly by the equation

$$F(x,t) = c, \quad c \text{ is constant.}$$

We note that if there is a function F such that (2.8) holds, then

$$\frac{\partial g}{\partial t} = \frac{\partial^2 F}{\partial t \partial x} = \frac{\partial^2 F}{\partial x \partial t} = \frac{\partial h}{\partial t}.$$

Conversely, if the functions g and h are such that

$$\frac{\partial g}{\partial t} = \frac{\partial h}{\partial x}$$

then a function F satisfying (2.8) can be constructed by integration.

2.2.5 Integrating Factor

A differential equation can sometimes be made an exact equation by multiplying it by a function, called **integrating factor**. If $\mu = \mu(x,y)$ is to be an integrating factor for Eq. (2.7), then it has to satisfy the equation

$$\frac{\partial}{\partial t}(\mu g) = \frac{\partial}{\partial x}(\mu h),$$

or

$$\mu(\frac{\partial g}{\partial t} - \frac{\partial h}{\partial x}) + g\frac{\partial \mu}{\partial t} - h\frac{\partial \mu}{\partial x} = 0.$$

If

$$\frac{1}{g}(\frac{\partial g}{\partial t} - \frac{\partial h}{\partial x}) \text{ is a function } k(t) \text{ (of } t \text{ only),}$$

then we can find an integrating factor $\mu = \mu(t)$ by solving

$$\frac{1}{\mu}\frac{d\mu}{dt} = -k(t).$$

On the other hand if

$$\frac{1}{h}(\frac{\partial g}{\partial t} - \frac{\partial h}{\partial x}) \text{ is a function } m(x) \text{ (of } x \text{ only),}$$

then we can find an integrating factor $\mu = \mu(x)$ by solving

$$\frac{1}{\mu}\frac{d\mu}{dx} = m(x).$$

Example 2.5. Consider the equation

$$t(t-x)\frac{dx}{dt} + (3xt - x^2) = 0.$$

In this case we have

$$\frac{1}{g}(\frac{\partial g}{\partial t} - \frac{\partial h}{\partial x}) = \frac{1}{t} \quad \text{and } \mu(t) = t.$$

Multiplying the differential equation by t we obtain an exact differential equation

$$(t^3 - tx)\frac{dx}{dt} + (3xt^2 - \frac{1}{2}x^2) = 0$$

with $F(x,t)$ such that

$$\frac{\partial F}{\partial x} = t^3 - tx, \quad \frac{\partial F}{\partial t} = 3xt^2 - \frac{1}{2}x^2,$$

namely

$$F(x,t) = t^3 x - \frac{1}{2}tx^2.$$

Hence the solution of the differential equation is

$$t^3 x - \frac{1}{2}tx^2 = c, \quad c \text{ is constant.}$$

We have actually already encountered an integrating factor for the linear equation (2.3), namely, e^P. Indeed, after multiplying both sides of Eq. (2.3) by e^P we obtain an exact equation with

$$F(x,t) = e^{P(t)}x(t) - \int_0^t e^{P(s)}g(s)ds.$$

2.2.6 Existence of Solutions

So far we have shown how to solve explicitly some classes of differential equations. For general functions $f(x,t)$ the initial value problem (2.1)–(2.2) cannot be solved explicitly, but it can always be solved numerically, as will be shown in the numerical sections. The following theorem asserts that the initial value problem (2.1)–(2.2) has a unique solution.

Theorem 2.1. *Let $f(x,t)$ be a continuously differentiable function in a domain which contains a point (x_0,t_0). Then the initial value problem (2.1)–(2.2) has a unique solution $x = x(t)$ for t in some interval which contains the point $t = t_0$.*

We shall be particularly interested in differential equation (2.1) where f is independent of t, namely,

$$\frac{dx}{dt} = f(x),   \tag{2.9}$$

and $f(x)$ is continuously differentiable for all x. In this case Theorem 2.1 can be extended as follows:

Theorem 2.2. *The solution of the initial value problem (2.9), (2.2) exists for all positive t as long as $x(t)$ remains bounded.*

The proof of Theorems 2.1 and 2.2 can be found, for instance, in Reference [1].

Example 2.6. Consider the system

$$\frac{dx}{dt} = x^\alpha, \quad x(0) = 1,   \tag{2.10}$$

where $0 < \alpha < \infty$. Rewriting the differential equation in the form

$$x^{-\alpha}dx = dt,$$

and we integrate it (note that the equation has separable variables) and use the initial condition to obtain

$$\frac{x^{1-\alpha}}{1-\alpha} = t + \frac{1}{1-\alpha},$$

or

$$x(t) = [(1-\alpha)t + 1]^{\frac{1}{1-\alpha}}.$$

If $0 < \alpha < 1$ then the solution exists for all $t > 0$ and $x(t) \to \infty$ as $t \to \infty$. If, however, $\alpha > 1$ then as t increases to $1/(\alpha - 1)$ the solution $x(t)$ increases to ∞, so the solution exists only for $t < 1/(\alpha - 1)$.

The solution of (2.9), (2.2) can also be continued to $t < 0$, but again only as long as $x(t)$ remains bounded. One often refers to a solution of (2.9), $x(t)$ for $0 \le t < \infty$, as a **trajectory**.

2.2.7 Differential Inequalities

We shall encounter in this book differential inequalities of the form

$$\frac{dx}{dt} + \mu x \le b \text{ for } t > 0, \tag{2.11}$$

or

$$\frac{dx}{dt} + \mu x \ge b \text{ for } t > 0, \tag{2.12}$$

and we shall need to determine the behavior of $x(t)$ as $t \to \infty$. Consider first the inequality (2.11). Multiplying both sides by $e^{\mu t}$ (note that $e^{\mu t}$ is always positive for real μ) we get

$$\frac{d}{dt}(e^{\mu t} x(t)) \le b e^{\mu t}$$

so that, by integration from 0 to t,

$$e^{\mu t} x(t) - x(0) < \frac{b}{\mu}(e^{\mu t} - 1)$$

or

$$x(t) \le e^{-\mu t}\left(x(0) - \frac{b}{\mu}\right) + \frac{b}{\mu}.$$

We conclude that if $x(0) \le \frac{b}{\mu}$ then $x(t) \le \frac{b}{\mu}$ for all $t > 0$. If however $x(0) > \frac{b}{\mu}$ then, for any small $\varepsilon > 0$,

$$x(t) < \frac{b}{\mu} + \varepsilon \tag{2.13}$$

if t is large enough, so that

$$e^{-\mu t}\left(x(0) - \frac{b}{\mu}\right) < \varepsilon.$$

Similarly, from the inequality (2.12) we can deduce that for any small $\varepsilon > 0$

$$x(t) > \frac{b}{\mu} - \varepsilon \tag{2.14}$$

if t is large enough. For later references we state:

Theorem 2.3. *If a function $x(t)$ satisfies the differential inequality (2.11), then, for any small $\varepsilon > 0$, (2.13) holds for all t sufficiently large. Similarly, if a function $x(t)$ satisfies the inequality (2.12), then, for any small $\varepsilon > 0$, (2.14) holds for all t sufficiently large.*

2.3 Equilibrium and Stability

If x_0 is a point such that $f(x_0) = 0$, then the unique solution of (2.9), (2.2) is clearly $x(t) \equiv x_0$. Such a point x_0 is called an **equilibrium point**, a **steady state**, or a **stationary point**. By Taylor's formula,

$$f(x) = f(x_0) + f'(x_0)(x - x_0) + (x - x_0)\varepsilon(x - x_0),$$

where $\varepsilon(x - x_0) \to 0$ if $x \to x_0$.

Suppose x_0 is an equilibrium point such that $f'(x_0) < 0$. Setting $y = x - x_0$ and by using Eq. (2.9), we then have

$$\frac{dy}{dt} = f'(x_0)y + y\varepsilon(y),$$

where $\varepsilon(y) \to 0$ if $y \to 0$.

If $|y|$ is small enough so that $|\varepsilon(y)| < \frac{1}{2}|f'(x_0)|$, then, for $y > 0$,

$$\frac{dy}{dt} < f'(x_0)y + \frac{1}{2}|f'(x_0)|y = f'(x_0)y - \frac{1}{2}f'(x_0)y = \frac{1}{2}f'(x_0)y,$$

so that

$$\frac{dy}{dt} < 0 \quad \text{if} \quad y > 0$$

and $y = y(t)$ is decreasing toward $y = 0$. Similarly

$$\frac{dy}{dt} > 0 \quad \text{if} \quad y < 0,$$

so that $y = y(t)$ is increasing toward $y = 0$.

Hence when $f'(x_0) < 0$, the solution $x(t)$, starting near x_0, moves toward x_0 as t increases; in fact, $x(t) \to x_0$ as $t \to \infty$. We therefore call x_0 a **stable equilibrium** (or more precisely **asymptotically stable equilibrium**). Similarly, if

$$f'(x_0) > 0$$

then solutions initiating near x_0 move away from x_0, as long as they are within a small distance from x_0. We call such a point x_0 an **unstable** equilibrium.

A steady point x_0 is called **globally (asymptotically) stable** if $x(t) \to x_0$ for any trajectory $x(t)$ whose initial value $x(0)$ is not a steady point.

2.4 Growth Models

We need to develop a mathematical model describing the growth of bacteria population. The density x of bacteria is defined as the number of bacteria per unit volume.
If the bacteria grow at a fixed rate r, then

$$x(t + \Delta t) - x(t) = rx(t)\Delta t,$$

or

$$\frac{x(t + \Delta t) - x(t)}{\Delta t} = rx(t),$$

and, taking $\Delta t \to 0$, we get

$$\frac{dx}{dt} = rx. \tag{2.15}$$

The explicit formula for the growth of x is then

$$x(t) = x(0)\, e^{rt}.$$

The **doubling time** T is defined by $x(T) = 2x(0)$, that is, the time for the bacteria
to double in number, and it is given by

$$2 = e^{rT}, \quad \text{or} \quad T = \frac{\ln 2}{r}.$$

If a colony of bacteria, or other microorganism, is dying at rate s, then its density x
satisfies

$$\frac{dx}{dt} = -sx, \tag{2.16}$$

and

$$x(t) = x(0)e^{-st}.$$

The population density is halved at time $\bar{T}$, called the **half-life**, given by

$$\bar{T} = \frac{\ln 2}{s}.$$

When bacteria are confined within a bounded chamber, they cannot grow expo-
nentially forever, by following (2.15). There is going to be a **carrying capacity** B
of the medium which the bacterial density cannot exceed. This situation is modeled
by replacing the exponential growth (2.15) by the **logistic growth**

$$\frac{dx}{dt} = rx(1 - \frac{x}{B}). \tag{2.17}$$

The solution of (2.17) with an initial condition

$$x(0) = x_0$$

is given by

$$x(t) = \frac{B}{1 + (\frac{B}{x_0} - 1)e^{-rt}}.$$ (2.18)

Indeed, to derive (2.18), we rewrite (2.17) in the form

$$\frac{dx}{x(1 - \frac{x}{B})} = rdt,$$

or

$$(\frac{1}{x} + \frac{1}{B}\frac{1}{1 - \frac{x}{B}})dx = rdt,$$

and integrate to obtain

$$\ln x - \ln(1 - \frac{x}{B}) = rt + const.$$

Then

$$\frac{x}{1 - \frac{x}{B}} = Ce^{rt},$$

and solving for x we get

$$x(t) = \frac{Ce^{rt}}{1 + \frac{C}{B}e^{rt}} = \frac{B}{1 + \frac{B}{C}e^{-rt}}.$$ (2.19)

Substituting $t = 0$, $x(0) = x_0$, we find the value of C:

$$1 + \frac{B}{C} = \frac{B}{x_0}, \quad \text{or} \quad C = \frac{x_0}{1 - \frac{x_0}{B}}.$$

Substituting C into Eq. (2.19), we obtain the formula (2.18) for the solution of Eq. (2.17). In the logistic growth equation (2.17) the point $x = B$ is a stable equilibrium. From (2.18) we see that $x = B$ is also a globally asymptotically stable equilibrium, since, for any initial value $x(0) = x_0$, $x(t) \to B$ as $t \to \infty$.

2.5 Modeling the Chemostat

Figure 2.2 shows a schematics of a chemostat with a stock of nutrient C_0 pumped into the chamber of the bacterial culture. We assume that the chemostat chamber is well stirred so that the nutrient concentration is constant at each time t. We then model the bacterial growth by the logistic equation (2.17), where r depends on the constant nutrient concentration C_0. If we denote by s the rate of the bacterial outflow from the chamber, then the balance between growth and outflow is given by

$$\frac{dx}{dt} = rx(1 - \frac{x}{B}) - sx.$$ (2.20)

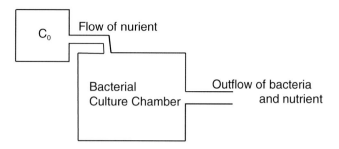

Fig. 2.2: The chemostat device.

We shall denote by $[X]$ the dimension of any quantity X. Then,

$$[x] = \frac{\text{number}}{\text{volume}}, \quad [B] = \frac{\text{number}}{\text{volume}},$$

$$[r] = \frac{1}{\text{time}}, \quad [s] = \frac{1}{\text{time}}.$$

There are two equilibrium points to (2.20), namely, $x = 0$, and $x = (1 - \frac{s}{r})B$. Note that if the outflow rate is less than the growth rate of the bacteria, that is, if $s < r$, then $x = 0$ is an unstable equilibrium, whereas $x = (1 - \frac{s}{r})B$ is a stable equilibrium. If $s > r$, then $x = 0$ is a stable equilibrium, whereas the equilibrium point $x = (1 - \frac{s}{r})B$ is not biologically relevant since it is negative.

Consider the case $s < r$ and $x(0) < (1 - \frac{s}{r})B$. Since $(1 - \frac{s}{r})B$ is a stable equilibrium, if $x(0)$ is near $(1 - \frac{s}{r})B$, it will remain smaller than $(1 - \frac{s}{r})B$ and will converge to it as $t \to \infty$. We can actually solve $x(t)$ explicitly: writing

$$\frac{1}{rx(1 - \frac{x}{B}) - sx} = \frac{1}{r-s}\left(\frac{1}{x} + \frac{r/B}{(r-s) - rx/B}\right)$$

we have

$$\frac{1}{r-s}\left[\frac{dx}{x} + \frac{r/B}{(r-s) - rx/B}dx\right] = dt.$$

By integration

$$\frac{1}{r-s}[\ln x - \ln((r-s) - rx/B)] = t + const,$$

or

$$\frac{x}{(r-s) - rx/B} = ce^{(r-s)t} \quad (c \text{ is constant}).$$

Hence

$$(\frac{1}{c}e^{-(r-s)t} + \frac{r}{B})x = r - s,$$

or

$$x(t) = \frac{r-s}{\frac{r}{B} + \frac{1}{c}e^{-(r-s)t}}.$$ (2.21)

We see that $x(t) \to (1 - \frac{s}{r})B$ as $t \to \infty$, whenever $x(0) < (1 - \frac{s}{r})B$. Note that the formula (2.21) is valid also when $x(0) > (1 - \frac{s}{r})B$ and that c is determined by

$$x(0) = \frac{r-s}{\frac{r}{B} + \frac{1}{c}}, \quad \text{or} \quad \frac{1}{c} = \frac{r-s}{x(0)} - \frac{r}{B}.$$

The chemostat operator would like to adjust the outflow rate s so as to get the largest output of bacteria. The mathematical model we developed can determine the optimal rate. Indeed, at steady state the outflow rate s is to be multiplied by the steady state of the bacteria, which is $x = (1 - \frac{s}{r})B$. The function $s(1 - \frac{s}{r})B$ takes its maximum at $s = \frac{r}{2}$, and with this output rate the maximum outflow per unit time is $\frac{1}{2}rB$.

Summary. The chemostat operates most efficiently when $s = \frac{r}{2}$, that is, when the outflow rate is half the inflow rate.

Problem 2.1. Find the general solution of the differential equations

(i) $\frac{dx}{dt} + x = 3e^t$;

(ii) $\frac{dx}{dt} = -2tx + t$;

(iii) $t\frac{dx}{dt} + \alpha x = t^2$, $\alpha > 0$.

Problem 2.2. Find the solution of the initial value problems

(i) $\frac{dx}{dt} - tx = t$, $x(0) = 2$;

(ii) $\frac{dx}{dt} - 3x = t + 2$, $x(0) = -1$.

Problem 2.3. Find the solution of the initial value problems

(i) $\frac{dx}{dt} = \frac{t}{x}$, $x(1) = 3$;

(ii) $\frac{dx}{dt} = \frac{1+x^2}{xt}$, $x(1) = 2$.

Problem 2.4. Solve the equation

$$\frac{dx}{dt} = \frac{x+4t}{x+t} \quad \text{with } x(0) = 3.$$

Problem 2.5. Find the solution of

$$\left(2xt + \frac{1}{x}\right)\frac{dx}{dt} = x^2, \quad x(3) = 1.$$

Problem 2.6. Find the solution of

$$(3x^3 + xt^2)\frac{dx}{dt} + 2x^2t = 0, \quad x(2) = 8.$$

Problem 2.7. Consider the equation

$$\frac{dx}{dt} = x(x-a)(x-2), \quad 0 < a < 2.$$

It has three steady points, $x = 0$, $x = 2$, and $x = a$. Determine which of them is stable.

Problem 2.8. Consider the equation

$$\frac{dx}{dt} = (x-a)(2-x), \qquad x(0) < a,$$

where $a < 2$. Find the solution explicitly in either the forms $t = t(x)$ or $x = x(t)$, and use it to prove the following:

(i) If $x(0) > a$ then the solution exists for all $t > 0$ and $x(t) \to 2$ as $t \to \infty$;

(ii) If $x(0) < a$ then the solution exists for $t < T$, where $T = \frac{1}{2-a} \ln|\frac{2-x(0)}{a-x(0)}|$, and $x(t) \to -\infty$ as $t \to T$.

2.6 Numerical Simulations – Introduction to MATLAB

MATLAB is a software developed by The MathWorks, Inc., and it is widely used in science and engineering. MATLAB is a high-level language and interactive environment for numerical computation, symbolic calculation, and visualization. It is also known for its easy handling of matrices and vectors. To access this software, in many universities, students can install licensed MATLAB software (you can request from the IT department in your school), and individual licenses can also be purchased through MathWorks website.

We will refer the readers to MathWorks' website for details of installation and launch of the software. In this chapter, we will introduce some basics of MATLAB and prompt to solving an ODE problem with MATLAB. The codes and explanations about MATLAB are based on the version of MATLAB R2014b.

The introduction here is elementary and not comprehensive, but it will give the readers the basic idea of how MATLAB operates and how to use this software to solve our models. We strongly encourage the readers to practice along when reading through numerical sections in this book.

2.6.1 Scalar Calculations

Once we launch MATLAB, the default window will have several compartments: a panel with function buttons, and main columns "Current folder," "Command Window," and "Workspace." We can change to the directory that we would like to work in, and the corresponding folders and subfolders will show in the "Current

Folder" part. The "Command Window" is for us to enter commands and do some calculations, and the "Workspace" will save the variables that have been used in our calculations.

MATLAB can do basic calculations as in regular calculators. MATLAB recognizes the usual arithmetic operation: + (addition), − (subtraction), * (multiplication), / (division), ^ (power). In the Command Window, we will see the prompt sign (>>), and we can type after prompt sign and press enter, which will give us the result of calculation. In the following, we show the MATLAB commands in teletype font. For example,

```
>> (5*2+3.5) / 5
ans =
2.7000
```

If we do not want to see the display of the answer, we can add a semicolon (;) after the command to suppress the display. We can also store the result into a variable that the user assigns, for example:

```
>> x = (5*2+3.5) / 5
x =
2.7000
```

If now we check the Workspace column, we will see that 'x' is stored and the value is also shown in that column. If we did not specify the name of the variable, the result will be stored in 'ans' in the Workspace. It is worth noting that a valid variable name starts with a letter, followed by letters, digits, or underscores. MATLAB is case sensitive, so B and b are not the same variable. We should avoid creating variable names that conflict with function names (functions will be introduced later).

MATLAB recognizes different types of numbers: (1) integer (example: 112, −2185); (2) real number (example: 2.452, −100.448); (3) complex (example: $-0.11 + 4.4i, i = \sqrt{-1}$); (4) Inf (infinity); (5) NaN (not a number).

All the calculations in MATLAB are done in double precision, which means that the numbers are accurate up to about 15 significant figures. However, we may not see that many digits on the display window, and this is because the default output format is to display 4 decimal places. If you type `format long` in the command window followed by pressing enter, for all the numbers shown in the command window, you will see the full display of all the digits. The command `format short` will switch back to display of 4 decimal places. To know about more format, type `help format`. In general, this `help` command is very useful when we would like to know how to use a command or a function; we simply type `help xx`, in which `xx` is the command of interest.

MATLAB has some built-in trigonometric functions and elementary functions. We choose some commonly used ones to list in Table 2.1.

When we code, it is usually important to make comments in the codes. These comments explain what the commands are for, so that the codes are easier to read later. In MATLAB, we use the percentage sign (%) to begin a comment, and MATLAB will take all the characters after (%) as comments and those will not be executed. For example:

```
>> y = (5*2+3.5)/5^2 % store the result in variable y,
and show the result on the screen.
```

Table 2.1: Commonly used MATLAB built-in functions. One can substitute 'x' in the table by numbers or other variables.

MATLAB build-in functions	descriptions
abs(x)	absolute value of x
sqrt(x)	square root of x
sin(x)	sine of x in radians
sind(x)	sine of x in degrees
cos(x)	cosine of x in radians
cosd(x)	cosine of x in degrees
tan(x)	tangent of x in radians
cot(x)	cotangent of x in radians
sec(x)	secant of x in radians
csc(x)	cosecant of x in radians
asin(x)	inverse sine of x in radians
acos(x)	inverse cosine of x in radians
atan(x)	inverse tangent of x in radians
sinh(x)	hyperbolic sine of x in radians
cosh(x)	hyperbolic cosine of x in radians
exp(x)	exponential of x
log(x)	natural logarithm of x
log2(x)	base 2 logarithm of x
log10(x)	base 10 logarithm of x
ceil(x)	round x toward infinity
floor(x)	round x toward minus infinity
round(x)	round x to the nearest integer

If the operation is too long, one can use '...' to extend the command to the next line, for example:

```
>> z = 10*sin(pi/3)*...
>> sin(pi^2/4)
```

A convenient way to record the commands we are typing is to use 'diary FILE-NAME', for example:

```
>> diary myfile
>> x = sqrt(5);
>> y = exp(x);
>> diary off
```

In the same directory, if you open the file 'myfile,' we will see the records of commands and outputs. We can turn the diary back on by using 'diary on.'

2.6.2 Vector and Matrix Operations

In previous examples, we have discussed how to use MATLAB to do the usual scalar calculations. In fact, MATLAB is very powerful when it comes to calculations of vectors and matrices, and it is a vector oriented programming language. For this reason, we should maximize the use of vector and matrix operations in our codes.

In the previous section, variables were used to store scalars. Here we show that they can also be used to store vectors. The following is an example to assign a vector to a variable:

```
>> s = [1 3 5 2]; % note the use of [], and the spaces
between the numbers; one can also use comma (,) to
separate the numbers
>> t = 2*s + 1 % 1 will be added to all the entries of 2*s
t =
3 7 11 5
```

In the above example, MATLAB uses [] to establish a row vector [1 3 5 2] and stores it in the variable s, and does an operation on it to make a new row vector [3 7 11 5] and stores it in the variable t. To extract one element from the vector or part of the vector to do operations, we type:

```
>> t(3) % display the third entry of vector t
ans =
  11
>> t(3) = 2 % assign another value to the third entry of
vector t
t =
  3 7 2 5
>> 2*t - 5*s
ans =
  1 -1 -21 0
```

As we have learned in linear algebra, in order to add or subtract, two vectors need to have the same length.

```
>> a = [1 2 3]; b = [5 6];
>> a + b
Error using +
Matrix dimensions must agree.
```

The above message means we have inconsistent matrix or vector dimensions, so we need to go back to check the dimensions of our matrices or vectors. Although we cannot add or subtract a and b, we can combine them to form a new vector, for example,

```
>> cd = [-b, 3*a]
cd =
 -5 -6 3 6 9
```

Sometimes, we need vectors whose entries are part of an arithmetic sequence, a convenient way to define it is to use the colon notation:

```
>> 1:2:6 % this will generate a row vector, starting
at 1, ending at 6, with increment 2
ans =
  1 3 5
```

```
>> 3:10 % without specifying the increment, it will
be set as 1
ans =
 3 4 5 6 7 8 9 10
```

With this trick, we can easily extract a part of a vector, and do operations:

```
>> t(2:4) - 1 % this will be the same as typing
t([2 3 4])-1
ans =
 6 1 4
```

We have learned how to define and use row vectors. The operations for column vectors are similar. The only difference is that the entries of a column vector are separated by semicolon (;) or by making a new line.

```
>> cv = [-1; pi; exp(2)]
cv =
 1.0000
 3.1416
 7.3891
>> cv2 = [1
2
3]
cv2 =
 1
 2
 3
```

The row and column vectors can be transposed to become column and row vectors, respectively. The transpose of a vector or matrix is done by putting an apostrophe after the variable name.

```
>> cv', t'
ans =
 1.0000 3.1416 7.3891
ans =
 3
 7
 2
 5
```

Similar to creating vectors, an $m \times n$ matrix can be created by adding a semicolon (;) after the end of each row. As in row and column vectors, entries in a row are separated by spaces or commas, while rows are separated by using semicolons or by making a new line. For example:

```
>> A = [1 2 3 4; 5 6 7 8; 9 10 11 12]
A =
 1 2 3 4
 5 6 7 8
 9 10 11 12
```

We can extract or change any single entry in the matrix
```
>> A(2,3) = 5; % change the (2,3) entry of A to 5
```
or extract part of the matrix
```
>> B = A(2,1:3) % take the second row, the first
to third column, store as a new matrix B
>> B =
   5 6 7
```
We can combine matrices, as long as the dimensions are consistent.
```
>> A =[A B'] % transpose B, make it as the last column
vector and merge with A
A =
 1 2  3  4 5
 5 6  7  8 6
 9 10 11 12 7
```
We can extract the whole row or column by using semicolon
```
>> A(:,3) % note here ':' can be replaced by '1:end',
that is, 1 to end
A =
  3
  7
 11
>> A(1,:)
A =
 1 2 3 4 5
```
Then we can redefine, or delete a row or a column from a matrix A:
```
>> A(:,2) = [] % delete the second column of A
(: represents all the rows, [] is an empty vector)
>> A = [A; 4 3 2 1; 0 -1 -2 -3]; % adding the fourth and
fifth row in the matrix A
```
To find out the dimension of a matrix, we use the command "size."
```
>> size(A') % the output is [number of rows, number of
columns]
ans =
 4 5
```
To obtain the length of a vector, we use "length."
```
>>length(A(1,:))
ans =
 4
```
  There are some built-in matrix generating functions,
```
>> ones(2,3) % this generates a 2x3 matrix with ones
>> zeros(4,4) % this generates a 4x4 matrix with zeros
>> eye(5) % this generates a 5x5 identity matrix
>> diag([1 3 5]) % this generates a matrix with 1 3 5 on
its diagonal
```

Next, let us do matrix-matrix or matrix-vector multiplication. When we use * in the matrix operations, it will operate as the matrix-matrix multiplication in linear algebra. For example,

```
>> X = [1 2 3; 0 2 4]; Y = [5 2; 1 1; 10 7]; W = X*Y
W =
 37 25
 42 30
```

If we try

```
>> X*X
```

then we will see an error message about the matrix dimension, because an $m \times n$ matrix can only be multiplied by an $n \times k$ matrix. Sometimes we would like to perform component-by-component operations, but not matrix-matrix multiplications; for that purpose we need to use '.*' instead of '*'. The following commands will give different results:

```
>> W.* W % component-by-component operation
>> W * W % matrix-matrix multiplication
```

and we will find that X.*X works because it is a component-by-component operation. Note that the use of '.*' requires the two matrices to have the same size. This component-wise operation of matrices can also be used for division ('./') and exponents ('.^').

Problem 2.9. Try the following command to generate a vector **x**.

```
>> x = 0:0.01:2
```

What is the x you see on MATLAB? Then use the command

```
>> y = sin(x)
```

to generate another vector **y**, what is **y**?

Problem 2.10. Let $\mathbf{x} = [2, 5, 1, 6]$.

(a) Add 15 to each element. [Hint: $\mathbf{x} + 15$.]
(b) Add 3 to only the odd-indexed elements of **x**.
(c) Output the vector whose elements are squares of the corresponding elements of **x**. [Hint: .* or .^]

Problem 2.11. Let $\mathbf{x} = [3, 1, 6, 8]'$ and $\mathbf{y} = [2, 1, 3, 5]'$ (**x** and **y** are column vectors).

(a) Add **x** to **y**.
(b) Raise each element of **x** to the power specified by the corresponding element in **y**.
(c) Divide each element of **y** by the corresponding element in **x**. [Hint: ./]
(d) Multiply each element in **x** by the corresponding element in **y**, and call the result '**z**'. [Hint: .*]
(e) Add up the elements in **z** and assign the result to a variable called '**w**'. [Hint: w=sum(z).]
(f) Compute $(\sin \mathbf{x})' * \mathbf{y} - \mathbf{w}$.

2.6.3 Program Files

MATLAB program (script) files are essentially text files with a file extension '.m'.
We can start a new script file simply by clicking an icon on the MATLAB window
called 'New Script' (name may vary in different systems or versions). A program
file can contain a series of commands to be executed (scripts), or it can contain a
function that accepts input arguments and produces output. Let's open a new script
file and type the following in the file:

```
a = 3.5;
b = 1;
x = sin(a)-b;
```

Save this file as 'testscript.m' in our working directory. In the command window,
type

```
>> testscript
```

and press enter, and you will see

```
x =
 -1.3508
```

We can also execute the script file by directly clicking the 'Run' button on the script
file window.

Another way to obtain the result is to make the file a function file. Open another
script file, and name it 'fun1.m'. In the file, type and save

```
function x = fun1(a,b)
x = sin(a)-b;
```

In the command window, type

```
>> fun1(3.5,1)
```

and you can see that it produces the same result as before. We can try more sets of
(a,b)

```
>> a1 = fun1(2,-1)
>> a2 = fun1(-2.4,10)
```

2.6.4 Numerical Algorithms for Solving Ordinary Differential Equations

Most of the time, the solution of an ordinary differential equation problem (2.1)
does not have a closed-form solution. In this case, one looks for numerical solutions
that approximate the exact solution. Since numerical solutions are just approximations, it is also important to understand the accuracy and robustness of the numerical
method.

Suppose the initial value problem is

$$\frac{dx}{dt} = f(x,t), \quad t \geq t_0, \quad x(t_0) = x_0. \tag{2.22}$$

Let t_n be some time point with $t_n \geq t_0$, then by integrating the equation from t_n to t, one gets

$$x(t) = x(t_n) + \int_{t_n}^{t} f(x,\tau)d\tau \approx x(t_n) + (t - t_n)f(x(t_n),t_n). \tag{2.23}$$

The approximation of the integral in (2.23) is good as long as t is sufficiently close to t_n. Suppose we would like to compute the solution of (2.22) at $t = T$, $T > t_0$. To get an approximate solution at time T, we can discretize the interval $[t_0, T]$ into N uniform subintervals $[t_n, t_{n+1}], n = 0,..,N-1$, with $t_N = T$ and $t_{n+1} - t_n = h = \frac{T}{N}$. We call h the step size. We will use lowercase x to denote the exact solution of (2.22) and capital X to denote the approximate solution.

Using the approximation in (2.23), we then define a numerical scheme by

$$X_{n+1} = X_n + hf(X(t_n),t_n). \quad n = 0, \cdots, N-1, \tag{2.24}$$

where X_n is the approximation of $x(t_n)$. This is called the **forward Euler Method**, named after Leonhard Euler (1707–1783). The error of this scheme is $O(h)$, which can be formally derived from the Taylor expansion. Hence, the smaller the time step size is, the more accurate the approximate solution will be. Generally, a numerical scheme is called k-th order accurate if the error is $O(h^k)$, where h is the discretization size. So Euler method is first order accurate. Although nowadays there are many high order accurate schemes to solve ordinary differential equations, Euler method is still a classical one when we first learn numerical methods. In MATLAB, we have some options of using Runge-Kutta methods [2] to solve ordinary differential equations, which will be introduced as follows.

Using MATLAB to Solve ODE

When solving problem (2.22) with MATLAB, we need to provide three pieces of information for the program:

1. the right-hand side function $f(x,t)$;
2. the initial condition $x(t_0) = x_0$;
3. the integration interval $[t_0, T]$.

The first step is to define functions in MATLAB. Recall that we introduced in the previous subsection about using a function file to define a single function, which reads in arguments and produces outputs. Another way is to use 'function handle,' a MATLAB value that provides a means of calling a function indirectly. For example, to define $f(x,t) = t - 2x$, we can type in the command window

```
>> f = @(x,t) t-2*x; %The @ operator constructs a
function handle for this function
>> f(3,1)
ans =
    -5
```

Problem 2.12. Try to use a script file to define the above function.

Now, to solve a simple ODE

$$\frac{dx}{dt} = t - 2x, \quad 0 \le t \le 2, \quad x(0) = 1,$$

we can type the following in the command window:

```
>> g = @(t,x)(t-2*x);
>> tspan = [0, 0.2]; % integrate the ODE from 0 to 0.2
>> x0 = 2; % the initial condition x(0) = 2
>> [t,x] = ode45(g,tspan,x0)
```

Note that the first argument in the function g is 't' and the second is 'x'; we have to keep this order (time is first, followed by other variables) when we define functions for MATLAB ODE solvers. The ODE solver we used is 'ode45,' a built-in Runge-Kutta solver in MATLAB.

Also, when we output variables t and x, we can see that t and x are column vectors. The vector t records the discrete time points in the MATLAB simulations, starting at 0 (the initial time) and ending at 0.2 (the final time). The vector x is of the same length as t, and the elements are the approximate solutions at time corresponding to elements in vector t (the first element of x is 2, which is the initial condition).

We can save the above commands in a script file so that we do not have to retype next time. A slightly different version is to use a function file to define the right-hand side function $f(x,t)$, see Algorithms 2.1 and 2.2 (run 'main_BacterialGrowth.m', and 'fun_BacterialGrowth.m' is to be called when 'ode45' is solving the ODE). A plot of x versus t will be shown by the 'plot' command.

Algorithm 2.1. Main script file to solve $dx/dt = t - 2x$ (main_BacterialGrowth.m)

```
%%% This code solves the ODE dx/dt=t-2x, 0<=t<=0.2 with x(t=0)=2

tspan = [0,0.2]; % integrate the ode from 0 to 0.2
x0 = 2;          % the initial condition x(0) = 2
[t,x] = ode45('fun_BacterialGrowth',tspan,x0);
plot(t,x)
```

Algorithm 2.2. fun_BacterialGrowth.m

```
%%% This function will be called by main_BacteriaGrowth.m
function dx = fun_BacterialGrowth(t,x)
dx = t - 2*x;
```

Problem 2.13. Write a code to solve the ODE (refer to Eq. (2.17))

$$\frac{dx}{dt} = x\left(1 - \frac{x}{2}\right), \quad 0 \le t \le 10,$$

with initial condition $x(0) = 0.5$.

(a) Run the code and get the two column vectors of discrete time points and the corresponding approximate solutions.
(b) Use the vector of time points to compute a vector containing the exact solution at those time points. [Hint: refer to formula (2.18); exponential of x is 'exp(x)' in MATLAB.]
(c) Compute the absolute value of the difference between the approximate and exact solutions.
(d) Plot the numerical solution and the exact solution on the same figure with different markers and different colors (refer to the numerical section of Chapter 3 for plotting).

Problem 2.14. Solve the equation in Problem 2.8 with $a = 1$ numerically in the form $x = x(t)$ when (i) $x(0) = \frac{1}{2}$, (ii) $x(0) = \frac{3}{2}$. For (i), plot x for the time interval when finite solution exists (starting from 0); for (ii), plot x for $0 \le t \le 10$.

Chapter 3
System of Two Linear Differential Equations

In Chapter 5 we shall model the interaction between predator y and prey x by a system of two differential equations: the differential equation for x will involve the predator y and the differential equation for y will involve the prey x. In order to study this model, as well as other models that will appear in subsequent chapters, we need to develop some basic theory for a system of two differential equations of order 1,

$$\frac{dx}{dt} = f(x,y), \quad \frac{dy}{dt} = g(x,y).$$

The functions $f(x,y)$ and $g(x,y)$ will generally be nonlinear functions. We shall develop the theory in two stages: The first stage to be taken up in this chapter deals with the special case where f and g are linear functions, and the second stage, to be taken up in Chapter 4, will extend the theory to nonlinear functions f and g. Before we start, with a linear system of two equations, however, it will be instructive to consider one linear differential equations of the second order.

3.1 Second Order Linear Differential Equations

Consider a second order differential equation

$$a\frac{d^2x}{dt^2} + b\frac{dx}{dt} + cx = 0, \tag{3.1}$$

where a, b, c are real constants and $a \neq 0$. The general solution is

$$x(t) = c_1 e^{\lambda_1 t} + c_2 e^{\lambda_2 t}, \quad c_1, c_2 \text{ are constants,} \tag{3.2}$$

Electronic supplementary material The online version of this chapter (doi: 10.1007/978-3-319-29638-8_3) contains supplementary material, which is available to authorized users.

© Springer International Publishing Switzerland 2016
C.-S. Chou, A. Friedman, *Introduction to Mathematical Biology*,
Springer Undergraduate Texts in Mathematics and Technology,
DOI 10.1007/978-3-319-29638-8_3

where λ_1, λ_2 are the solutions of the quadratic equation

$$a\lambda^2 + b\lambda + c = 0,$$

namely,

$$\lambda_{1,2} = \frac{1}{2a}(-b \pm \sqrt{b^2 - 4ac}) \tag{3.3}$$

provided $\lambda_1 \neq \lambda_2$. If $\lambda_1 = \lambda_2 = -\frac{b}{2a}$, then, as easily seen, $te^{\lambda_1 t}$ is another solution of (3.1), and the general solution of (3.1) is

$$x(t) = c_1 e^{\lambda_1 t} + c_2 t e^{\lambda_1 t}. \tag{3.4}$$

We can use the general solution to solve Eq. (3.1) subject to initial conditions

$$x(0) = \alpha, \quad x'(0) = \beta. \tag{3.5}$$

Indeed, if $\lambda_1 \neq \lambda_2$ then c_1 and c_2 are uniquely determined by solving the equations

$$c_1 + c_2 = \alpha, \quad \lambda_1 c_1 + \lambda_2 c_2 = \beta,$$

and the solution is

$$c_1 = \frac{\alpha \lambda_2 - \beta}{\lambda_2 - \lambda_1}, \quad c_2 = \frac{\beta - \alpha \lambda_1}{\lambda_2 - \lambda_1}.$$

If $\lambda_1 = \lambda_2$ then

$$c_1 = \alpha \quad \text{and} \quad c_2 = \beta - \lambda_1 \alpha.$$

If $b^2 - 4ac$ is negative, then λ_1 and λ_2 are complex numbers,

$$\lambda_{1,2} = \frac{1}{2a}(-b \pm i\sqrt{4ac - b^2}) = \mu \pm i\nu \tag{3.6}$$

and

$$e^{\lambda_{1,2} t} = e^{\mu t}(\cos \nu t \pm i \sin \nu t).$$

Then the general real-valued solution can be written in the form

$$x(t) = c_1 e^{\mu t} \cos \nu t + c_2 e^{\mu t} \sin \nu t. \tag{3.7}$$

If we set

$$y = \frac{dx}{dt},$$

then Eq. (3.1) can be written as a system of linear differential equations,

$$\frac{dx}{dt} = y, \tag{3.8}$$

$$\frac{dy}{dt} = -\frac{c}{a}x - \frac{b}{a}y.$$

From Eq. (3.4) we deduce that if $\lambda_1 \neq \lambda_2$ then the general solution of this system is

$$\begin{pmatrix} x(t) \\ y(t) \end{pmatrix} = c_1 \begin{pmatrix} e^{\lambda_1 t} \\ \lambda_1 e^{\lambda_1 t} \end{pmatrix} + c_2 \begin{pmatrix} e^{\lambda_2 t} \\ \lambda_2 e^{\lambda_2 t} \end{pmatrix}.$$

If $\lambda_{1,2}$ are complex numbers, then, by (3.7) the real-valued general solution is

$$\begin{pmatrix} x(t) \\ y(t) \end{pmatrix} = c_1 \begin{pmatrix} e^{\mu t} \cos \nu t \\ e^{\mu t} (\mu \cos \nu t - \nu \sin \nu t) \end{pmatrix} + c_2 \begin{pmatrix} e^{\mu t} \sin \nu t \\ e^{\mu t} (\mu \sin \nu t + \nu \cos \nu t) \end{pmatrix}.$$

3.2 Linear Systems

We shall need some basic facts from Linear Algebra. We first recall that, for any matrix

$$A = \begin{pmatrix} \alpha_{11} & \alpha_{12} \\ \alpha_{21} & \alpha_{22} \end{pmatrix},$$

one defines the determinant of A by

$$\det A = \begin{vmatrix} \alpha_{11} & \alpha_{12} \\ \alpha_{21} & \alpha_{22} \end{vmatrix} = \alpha_{11}\alpha_{22} - \alpha_{12}\alpha_{22}.$$

Consider a linear system

$$\begin{aligned} \alpha_{11}x_1 + \alpha_{12}x_2 &= b_1, \\ \alpha_{21}x_1 + \alpha_{22}x_2 &= b_2, \end{aligned} \tag{3.9}$$

where $A = (\alpha_{ij})$ is a given matrix. We wish to have a unique solution (x_1, x_2) for any prescribed vector (b_1, b_2). The following theorems give a complete answer.

Theorem 3.1. *If $\det A \neq 0$ then for any vector (b_1, b_2) there exists a unique solution of the system (3.9), and it is given by*

$$x_1 = \frac{\begin{vmatrix} b_1 & \alpha_{12} \\ b_2 & \alpha_{22} \end{vmatrix}}{\det A}, \quad x_2 = \frac{\begin{vmatrix} \alpha_{11} & b_1 \\ \alpha_{21} & b_2 \end{vmatrix}}{\det A};$$

in particular, if $b_1 = b_2 = 0$ then the unique solution is $x_1 = x_2 = 0$.

Theorem 3.2. *If $\det A = 0$ then the homogeneous system*

$$\begin{aligned} \alpha_{11}x_1 + \alpha_{12}x_2 &= 0 \\ \alpha_{21}x_1 + \alpha_{22}x_2 &= 0 \end{aligned}$$

has nonzero solutions; hence, if the only solution of the homogeneous system is $x_1 = x_2 = 0$ then $\det A \neq 0$.

In this section we consider a general system of differential equations with constant coefficients

$$\frac{dx_1}{dt} = a_{11}x_1 + a_{12}x_2,$$
$$\frac{dx_2}{dt} = a_{21}x_1 + a_{22}x_2. \tag{3.10}$$

Motivated by the special case of system (3.8) we try to find a solution in the form

$$x_1 = v_1 e^{\lambda t}, \quad x_2 = v_2 e^{\lambda t},$$

where the coefficients v_1, v_2 are to be determined from the equations

$$a_{11}v_1 + a_{12}v_2 = v_1\lambda,$$
$$a_{21}v_1 + a_{22}v_2 = v_2\lambda.$$

We can rewrite this system in matrix form

$$\begin{pmatrix} a_{11} - \lambda & a_{12} \\ a_{21} & a_{22} - \lambda \end{pmatrix} \begin{pmatrix} v_1 \\ v_2 \end{pmatrix} = \begin{pmatrix} 0 \\ 0 \end{pmatrix}, \tag{3.11}$$

or $(A - \lambda I)\mathbf{v} = 0$, where

$$A = \begin{pmatrix} a_{11} & a_{12} \\ a_{21} & a_{22} \end{pmatrix}, \quad I = \begin{pmatrix} 1 & 0 \\ 0 & 1 \end{pmatrix}, \quad \mathbf{v} = \begin{pmatrix} v_1 \\ v_2 \end{pmatrix}.$$

By Theorems 3.1 and 3.2, a nonzero solution $\mathbf{v}$ exists if and only if λ satisfies the **characteristic equation**

$$\det(A - \lambda I) = 0. \tag{3.12}$$

A solution λ of (3.12) is called an **eigenvalue** of A and a corresponding vector $\mathbf{v}$ is called **eigenvector**. Eq. (3.12) can be written explicitly as

$$\lambda^2 - \lambda(a_{11} + a_{22}) + (a_{11}a_{22} - a_{12}a_{21}) = 0. \tag{3.13}$$

If the two eigenvalues λ_1, λ_2 are different, then the general solution of the system (3.9) is

$$\mathbf{x}(t) = c_1 \mathbf{w}_1 e^{\lambda_1 t} + c_2 \mathbf{w}_2 e^{\lambda_2 t}, \tag{3.14}$$

where $\mathbf{w}_1$ and $\mathbf{w}_2$ are eigenvectors corresponding to λ_1 and λ_2, respectively. More precisely:

Theorem 3.3. *If $\lambda_1 \neq \lambda_2$ then, for any initial values*

$$\mathbf{x}(0) = \mathbf{b}, \quad \text{where } \mathbf{b} = \begin{pmatrix} b_1 \\ b_2 \end{pmatrix}, \tag{3.15}$$

there is a unique solution of (3.10), (3.15) in the form (3.14).

Proof. We first claim that $\mathbf{w}_1, \mathbf{w}_2$ are linearly independent, that is,

$$\text{if } \alpha_1 \mathbf{w}_1 + \alpha_2 \mathbf{w}_2 = 0 \text{ then } \alpha_1 = \alpha_2 = 0.$$

Indeed this relation implies that

$$\alpha_1 \lambda_1 \mathbf{w}_1 + \alpha_2 \lambda_2 \mathbf{w}_2 = \alpha_1 A \mathbf{w}_1 + \alpha_2 A \mathbf{w}_2 = A(\alpha_1 \mathbf{w}_1 + \alpha_2 \mathbf{w}_2) = 0.$$

Since also $\alpha_1 \mathbf{w}_1 + \alpha_2 \mathbf{w}_2 = 0$, or $\alpha_1 \lambda_1 \mathbf{w}_1 = -\lambda_1 \alpha_2 \mathbf{w}_2$, we get, by subtraction,

$$\alpha_2 \lambda_2 \mathbf{w}_2 - \lambda_1 \alpha_2 \mathbf{w}_2 = 0, \quad \text{or } (\lambda_2 - \lambda_1) \alpha_2 \mathbf{w}_2 = 0.$$

If follows that $\alpha_2 = 0$, and then also $\alpha_1 = 0$.

Setting

$$\mathbf{w}_1 = \begin{pmatrix} v_{11} \\ v_{12} \end{pmatrix}, \quad \mathbf{w}_2 = \begin{pmatrix} v_{21} \\ v_{22} \end{pmatrix}$$

we conclude that

$$\text{if } \sum_{i=1}^{2} v_{ij} \alpha_i = 0 \text{ for } j = 1, 2, \text{ then } \alpha_1 = \alpha_2 = 0.$$

Hence, by Theorem 3.2, $\det(v_{ij}) \neq 0$. But then, by Theorem 3.1, for any (b_1, b_2) there is a unique solution (c_1, c_2) of the system

$$\sum_{i=1}^{2} v_{ij} c_i = b_i \quad (j = 1, 2),$$

and the function $\mathbf{x}(t)$ in (3.14) is then the solution asserted in the theorem.

Consider next the case where λ_1 is a complex number, $\lambda_1 = \mu + i\nu$. Then the components of the eigenvector $\mathbf{w}_1$ are also complex numbers. But we are interested only in real-valued solutions. So in order to construct real-valued solutions we write

$$\mathbf{w}_1 e^{\lambda_1 t} = \begin{pmatrix} v_{11} + i v_{12} \\ v_{21} + i v_{22} \end{pmatrix} e^{\mu t} (\cos \nu t + i \sin \nu t), \tag{3.16}$$

where v_{ij} are real numbers. We note that the complex conjugate of $\mathbf{w}_1 e^{\lambda_1 t}$ is also a solution of (3.10) and, hence, so are the real and imaginary parts of (3.16). It follows that

$$\mathbf{w}_1 = e^{\mu t} \begin{pmatrix} v_{11} \cos \nu t - v_{12} \sin \nu t \\ v_{21} \cos \nu t - v_{22} \sin \nu t \end{pmatrix} \quad \text{and} \quad \mathbf{w}_2 = e^{\mu t} \begin{pmatrix} v_{11} \sin \nu t + v_{12} \cos \nu t \\ v_{21} \sin \nu t + v_{22} \cos \nu t \end{pmatrix}$$
$$\tag{3.17}$$

are two solutions.

Theorem 3.4. *The two solutions $\mathbf{w}_1$ and $\mathbf{w}_2$ are linearly independent.*

Proof. Since $\mathbf{w}_1 \neq 0$, at least one of the numbers v_{ij} in (3.16) is not equal to zero. Suppose $v_{11} \neq 0$, the other cases can be treated similarly. If the assertion of the theorem is not true, then there are numbers γ_1, γ_2 such that

$$\gamma_1 \mathbf{w}_1 + \gamma_2 \mathbf{w}_2 = \mathbf{0} \tag{3.18}$$

and $\gamma_1 \neq 0$, or $\gamma_2 \neq 0$. We consider the first component in (3.18), and note that the sum of the coefficients of $\cos vt$ and the sum of the coefficients of $\sin vt$ must be equal to zero, so that, by (3.17),

$$v_{11}\gamma_1 + v_{12}\gamma_2 = 0,$$
$$-v_{12}\gamma_1 + v_{11}\gamma_2 = 0.$$

But since

$$\det \begin{pmatrix} v_{11} & v_{12} \\ -v_{12} & v_{11} \end{pmatrix} = v_{11}^2 + v_{12}^2 > 0,$$

we conclude, by Theorem 3.1, that $\gamma_1 = \gamma_2 = 0$, which is a contradiction to the assumption that $\gamma_1 \neq 0$ or $\gamma_2 \neq 0$.

From Theorem 3.4 it follows, as in the proof of Theorem 3.3, that any solution of (3.10) is a linear combination of the two solutions in (3.17).

By writing the roots λ_1, λ_2 of (3.13) in the form (3.3) or (3.6), we see that $Re\lambda_1 < 0$ and $Re\lambda_2 < 0$ if and only if

$$\begin{aligned} \text{trace of } A &\equiv a_{11} + a_{22} < 0, \\ \text{determinant of } A &\equiv a_{11}a_{22} - a_{12}a_{21} > 0. \end{aligned} \tag{3.19}$$

If $\lambda_1 = \lambda_2$, then in addition to a solution $\mathbf{w}_1 e^{\lambda_1 t}$ of the system (3.10) where $\mathbf{w}_1$ is an eigenvector of (3.11), there is another solution of the form $\mathbf{w}_1 t e^{\lambda_1 t} + \hat{\mathbf{w}}_2 e^{\lambda_1 t}$ where $\hat{\mathbf{w}}_2$ is an appropriate vector. Setting $\mathbf{w}_2 = \mathbf{w}_1 + \hat{\mathbf{w}}_2$, the general solution of the system (3.10) is

$$\mathbf{x}(t) = c_1 \mathbf{w}_1 t e^{\lambda_1 t} + c_2 \mathbf{w}_2 e^{\lambda_1 t}.$$

3.3 Equilibrium Points

We denote a variable point in the plane by $\mathbf{x} = (x_1, x_2)$. The point $\mathbf{x} = \mathbf{0}$ is called an **equilibrium** point of the system (3.10), since the solution $\mathbf{x}(t)$ with $\mathbf{x}(0) = \mathbf{0}$ is $\mathbf{x}(t) \equiv \mathbf{0}$. We define the **phase space** for the system (3.10) as the (x_1, x_2)-space, and we want to draw the portrait of the trajectories $(\mathbf{x}(t), t > 0)$ in this space near $\mathbf{x} = \mathbf{0}$, at least qualitatively. This can be done with the aid of the form (3.14) of the general solution. The portrait will depend on the eigenvalues λ_1, λ_2.

Figures 3.1(B) and 3.1(E) show that when both eigenvalues have negative real parts, all the trajectories converge to $\mathbf{x} = \mathbf{0}$; we say that $\mathbf{x} = \mathbf{0}$ is a **stable** equilibrium (or more precisely, **asymptotically stable** equilibrium). On the other hand, when

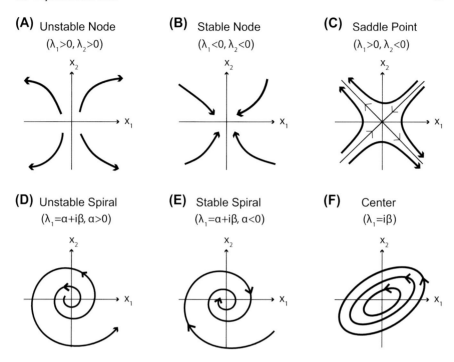

(A) Unstable Node
$(\lambda_1 > 0, \lambda_2 > 0)$

(B) Stable Node
$(\lambda_1 < 0, \lambda_2 < 0)$

(C) Saddle Point
$(\lambda_1 > 0, \lambda_2 < 0)$

(D) Unstable Spiral
$(\lambda_1 = \alpha + i\beta, \alpha > 0)$

(E) Stable Spiral
$(\lambda_1 = \alpha + i\beta, \alpha < 0)$

(F) Center
$(\lambda_1 = i\beta)$

Fig. 3.1: Phase portrait for system (3.9).

at least one of the eigenvalues has positive real part, there are always trajectories that go away from $\mathbf{x} = \mathbf{0}$ even if they start initially near $\mathbf{x} = \mathbf{0}$; we say that $\mathbf{x} = \mathbf{0}$ is an **unstable** equilibrium. Figures 3.1(A) and 3.1(D) are unstable, whereby all trajectories starting from $x(0) \neq 0$ go away to infinity, while in the case of the saddle point of Fig. 3.1(C), all but two trajectories go to infinity. In the (rather exceptional) case where both eigenvalues are pure imaginary numbers, the trajectories are all periodic, and $\mathbf{x} = \mathbf{0}$ is called a **center** point; see Fig. 3.1(F).

In order to solve an inhomogeneous linear equation

$$a\frac{d^2x}{dt^2} + b\frac{dx}{dt} + cx = f(t) \tag{3.20}$$

with a given function $f(t)$, we first need to find a special solution $\tilde{x}(t)$ and, then, the general solution is a sum of $\tilde{x}(t)$ and the general solution of the homogeneous equation. The same procedure applies to inhomogeneous linear systems.

Problem 3.1. Find the general solution of $x'' + x' - x = t^2$.

Problem 3.2. Find the solution of $x'' - 4x' + 3x = e^{-t}$ with $x(0) = \frac{1}{8}, x'(0) = \frac{1}{4}$.

Example 3.1. To find the general solution of

$$\frac{dx_1}{dt} = -x_1 + 6x_2$$

$$\frac{dx_2}{dt} = 2x_1 + 3x_2,$$

we substitute

$$x_1 = v_1 e^{\lambda t}, \quad x_2 = v_2 e^{\lambda t}$$

into the two differential equations. After canceling out $e^{\lambda t}$, we get two equations for v_1, v_2:

$$(-1 - \lambda)v_1 + 6v_2 = 0, \quad 2v_1 + (3 - \lambda)v_2 = 0.$$

A nonzero solution (v_1, v_2) exists if and only if

$$det \begin{pmatrix} -1 - \lambda & 6 \\ 2 & 3 - \lambda \end{pmatrix} = 0,$$

that is, if

$$(\lambda + 1)(\lambda - 3) - 12 = 0$$

or

$$\lambda^2 - 2\lambda - 15 = 0,$$

and the two solutions are $\lambda_1 = -3, \lambda_2 = 5$. If $\lambda = -3$ then the two equations for v_1, v_2 become identical,

$$2v_1 + 6v_2 = 0 \quad \text{and} \quad 2v_1 + 6v_2 = 0,$$

and we can take $v_1 = 3, v_2 = 1$. If $\lambda = 5$ then the two equations for v_1, v_2 are

$$-6v_1 + 6v_2 = 0 \quad \text{and} \quad 2v_1 - 2v_2 = 0,$$

and they are linearly dependent. So again a solution of one equation is also a solution of the other equation; we can take $v_1 = 1, v_2 = 1$. We conclude that

$$e^{-3t} \begin{pmatrix} 3 \\ -1 \end{pmatrix} \quad \text{and} \quad e^{5t} \begin{pmatrix} 1 \\ 1 \end{pmatrix}$$

are two solutions, and the general solution of the system is

$$c_1 e^{-3t} \begin{pmatrix} 3 \\ -1 \end{pmatrix} + c_2 e^{5t} \begin{pmatrix} 1 \\ 1 \end{pmatrix}$$

where c_1, c_2 are arbitrary constants.

Problem 3.3. Find the solution of

$$\frac{dx_1}{dt} = -2x_1 + 7x_2$$
$$\frac{dx_2}{dt} = 2x_1 + 3x_2.$$

Problem 3.4. Find the general solution of

$$\frac{dx_1}{dt} = x_1 - 2x_2$$
$$\frac{dx_2}{dt} = 2x_1 + x_2.$$

Problem 3.5. Find the general solution of

$$\frac{dx_1}{dt} = 3x_1 + 2x_2$$
$$\frac{dx_2}{dt} = -4x_1 - 3x_2.$$

Problem 3.6. Find the general solution of

$$\frac{dx_1}{dt} = -4x_1 + 6x_2$$
$$\frac{dx_2}{dt} = -3x_1 + 2x_2.$$

[Hint: Try to find solution of the form (3.17)].

In order to solve an inhomogeneous system

$$
\begin{aligned}
\frac{dx_1}{dt} &= a_{11}x_1 + a_{12}x_2 + f_1(t) \\
\frac{dx_2}{dt} &= a_{21}x_1 + a_{22}x_2 + f_2(t)
\end{aligned}
\tag{3.21}
$$

we need to find one special solution and add it to the general solution of the homogenous system (3.9).

Example 3.2. Find a special solution of

$$\frac{dx_1}{dt} = x_1 + 3x_2 - 4t^2 - 5$$
$$\frac{dx_2}{dt} = x_1 - x_2 + 1.$$

We try a polynomial solution

$$x_1 = a_1 t^2 + b_1 t + c_1, \quad x_2 = a_2 t^2 + b_2 t + c_2;$$

polynomials of degree higher than 2 are not needed since the equations for x_1, x_2 have polynomials in t of degree less than 2. Substituting the above polynomials into the differential equations, we get

$$2a_1t + b_1 = (a_1t^2 + b_1t + c_1) + 3(a_2t^2 + b_2t + c_2) - 4t^2 - 5$$

$$2a_2t + b_2 = (a_1t^2 + b_1t + c_1) - (a_2t^2 + b_2t + c_2) + 1.$$

Equating the coefficients of t^2 in each of the two equations, we find that

$$a_1 + 3a_2 - 4 = 0, \quad a_1 - a_2 = 0;$$

hence $a_1 = a_2 = 1$. Next we equate the coefficients of t, and, using the fact that $a_1 = a_2 = 1$, we get

$$2 = b_1 + 3b_2, \quad 2 = b_1 - b_2;$$

hence $b_1 = 2, b_2 = 0$. Finally, equating the remaining terms (which do not involve t^2 and t) we find (after using the fact that $b_1 = 2, b_2 = 0$) that

$$2 = c_1 + 3c_2 - 5, \quad 0 = c_1 - c_2 + 1;$$

hence $c_1 = 1, c_2 = 2$. Thus a special solution is $(1 + 2t + t^2, 2 + t^2)$.

Problem 3.7. Find a special solution of

$$\frac{dx_1}{dt} = -x_1 + x_2 - 2e^{-t}$$

$$\frac{dx_2}{dt} = 2x_1 - x_2 - e^{-t}.$$

and solve this system with

$$x_1(0) = 1, \quad x_2(0) = 2.$$

3.4 Numerical Simulations

3.4.1 Solving a Second Order ODE

In Chapter 2, we have simulated scalar first order ODEs with MATLAB. A natural question is whether we need additional MATLAB functions to simulate higher order equations. The answer is no. What we need to do is to convert higher order equations into systems of ODEs, and then we will simulate the ODE systems. Let's take a second order ODE as an example:

$$u''(t) + 16u'(t) + 192u(t) = 0$$

can be converted to

$$\begin{cases} x_1' = & x_2 \\ x_2' = -16x_2 - 192x_1 \end{cases}$$

by letting $x_1 = u$ and $x_2 = u'$. In general, a linear system of two first order ordinary differential equations has the form of system (3.21) which can also be written as

$$\frac{d}{dt}\begin{pmatrix} x_1 \\ x_2 \end{pmatrix} = \begin{pmatrix} a_{11} & a_{12} \\ a_{21} & a_{22} \end{pmatrix}\begin{pmatrix} x_1 \\ x_2 \end{pmatrix} + \begin{pmatrix} f_1(t) \\ f_2(t) \end{pmatrix}.$$

For example, given an ODE system

$$\frac{d}{dt}\begin{pmatrix} x_1 \\ x_2 \end{pmatrix} = \begin{pmatrix} 1 & 2 \\ 2 & 3 \end{pmatrix}\begin{pmatrix} x_1 \\ x_2 \end{pmatrix} + \begin{pmatrix} 0 \\ t^2 \end{pmatrix}, \quad 0 \le t \le 1,$$

with initial condition $\begin{pmatrix} x_1(0) \\ x_2(0) \end{pmatrix} = \begin{pmatrix} 2 \\ 3 \end{pmatrix}$, we can solve it with MATLAB, as shown in Algorithms 3.1 and 3.2.

Algorithm 3.1. Main file for solving a linear system (main_LinearDiffEqns.m)

```
%%% This code is to solve a linear system defined in
%%% fun_LinearDiffEqns.m

x_ini = [2,3]'; % the initial condition is a vector
[t,z] = ode45('fun_LinearDiffEqns', [0,1], x_ini)
% output t is a column vector,
% and the output z is a matrix containing two columns of same
% lengths as vector t
```

Algorithm 3.2. fun_LinearDiffEqns.m

```
%%% This function will be called by main_LinearDiffEqns.m
function dx = fun_LinearDiffEqns(t,x)

% the input is t (scalar) and x (a 2-by-1 column vector)
A = [1,2;2,3];
dx = A*x + [0; t^2]; % the output dx is a column vector
```

By running 'main_LinearDiffEqns.m', we have the output MATLAB variables t and z. Vector t is a column vector with components as the discrete time points that MATLAB uses in the simulation. Matrix z has two columns: the first column corresponds to the first variable x_1 and the second corresponds to the second variable x_2. Different rows in z represents the approximate values of x_1 and x_2 at the corresponding time points in t.

3.4.2 Plotting Figures

Suppose $x = [x_1, x_2, x_3, \cdots, x_n]$ is a vector representing sampling points on x−axis and $y = [y_1, y_2, y_3, \cdots, y_n]$ represents the corresponding function values of components of x (note that x and y must be of the same length), then to plot x versus y, one uses
```
>> plot(x,y)
```
To label the axis, we can use
```
>> xlabel('x'), ylabel('y')
```
This can also be done by a click on the figure window: 'Edit' → 'Axis properties' and then edit in the property editor.

One can also specify the color and marker by adding an option in the 'plot' function, for example,
```
>> plot(x,y,'r o') % this marks those point values by
red circles
```
If we would like to overlay two curves, x versus y and x versus z, where $z = [z_1, z_2, z_3, \cdots, z_n]$, we can use
```
      >> plot(x,y,'r',x,z,'b') % mark the first y(x) function
in red and the second z(x) in blue.
```
or we can use the following two commands
```
>> plot(x,y,'r'), hold on
>> plot(x,z,'b')
```
The 'hold on' command holds the first figure data and the second will be plotted on top of the first one. Without this command, the previous data in the figure will be overwritten.

Example 3.3. Let's type the following commands:
```
>> x = 0 : pi/100 : pi;
>> y = sin(x);
>> z = cos(x);
>> plot(x,y,'r',x,z,'g'), hold on
>> legend('y(x)','z(x)')
```
We can see the figure in Fig. 3.2. Note that using the command 'legend', we can add the legend to specify different curves; this can also be done by using the 'Insert' pull-down menu on the figure window. The font size and other properties of the figure can also be edited using 'Edit' pull-down menu in the figure window.

If we would like to plot $y(x)$ and $z(x)$ in one figure window but two separate plots, we can use 'subplot', for example, continuing the above commands:
```
>> close all % close all the figure windows
>> subplot(1,2,1), plot(x,y,'r')
>> subplot(1,2,2), plot(x,z,'b')
```

The figure will look like Fig. 3.3. The command 'subplot' defines an array of figures, with the first argument the number of rows and the second the number of columns. The third argument is from 1 to the product of the numbers of rows and columns, and the ordering runs through rows. If we rather separate the two figures into two figure windows and save them later in two files, we can type

```
>> close all % close all the figure windows
>> figure(1), plot(x,y,'r')
>> figure(2), plot(x,z,'b')
```

To save the figure, we can click 'Save as' in 'File' pull-down menu in the figure window and choose the format to save. We can save it as the usual figure format, or the Matlab figure file, with the extension '.fig', which we can later open with Matlab and edit directly.

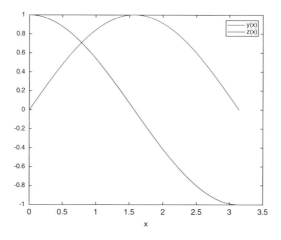

Fig. 3.2: Basic one-dimensional figure.

Problem 3.8. (a) Rewrite Problem 3.2 into a first order system. (b) Take the initial condition given in the problem, and the time interval $0 \leq t \leq 3$. Use MATLAB to solve the system you get in (a), and plot the two variables on the same figure.

Problem 3.9. (a) Solve $y'' - 5y' = 0$, $y(0) = 1$, $y'(0) = 2$ analytically; (b) solve the same problem with MATLAB, for $0 \leq t \leq 3$; (c) plot both the exact solutions obtained in (a) and the numerical solution in (b) with different colors.

Problem 3.10. Consider the system

$$\frac{dx_1}{dt} = x_1 - x_2$$

$$\frac{dx_2}{dt} = x_1 + x_2$$

with $x_1(0) = 1, x_2(0) = 5, 0 \leq t \leq 2$.

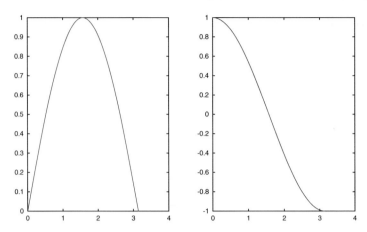

Fig. 3.3: Figure with subplots.

(a) Solve the system and write down the exact solution.
(b) Use MATLAB to solve the system.
(c) Plot on the same figure the exact solution and the numerical solution of $x_1(t)$, obtained in (a) and (b), respectively.

Chapter 4
Systems of Two Differential Equations

In Chapter 3, we considered linear differential system of the form (3.10). In this chapter we study general systems of two differential equations of the first order,

$$\frac{dx_1}{dt} = f_1(x_1,x_2), \quad \frac{dx_2}{dt} = f_2(x_1,x_2), \tag{4.1}$$

where $f_1(x_1,x_2), f_2(x_1,x_2)$ are any given functions, not necessarily linear. A point (a,b) such that

$$f_1(a,b) = 0, \quad f_2(a,b) = 0$$

is called an **equilibrium** point, a **stationary** point, or a **steady** point of the system (4.1). The x_1-**nullcline** of (4.1) is the curve consisting of points satisfying the equation

$$f_1(x_1,x_2) = 0.$$

Similarly, the x_2-**nullcline** is the curve defined by

$$f_2(x_1,x_2) = 0.$$

The equilibrium points of the system (4.1) are the points where the two nullclines intersect. To get an idea about how trajectories behave near a stationary point (a,b), we **linearize** the system.

We set

$$X_1 = x_1 - a, \quad X_2 = x_2 - b.$$

Then, by Taylor's formula, for $i = 1,2$

$$f_i(x_1,x_2) = f_i(a+X_1,b+X_2) = f_i(a,b) + \frac{\partial f_i}{\partial x_1}X_1 + \frac{\partial f_i}{\partial x_2}X_2 + \text{small terms},$$

Electronic supplementary material The online version of this chapter (doi: 10.1007/978-3-319-29638-8_4) contains supplementary material, which is available to authorized users.

where we abbreviate

$$\frac{\partial f_i}{\partial x_1} = \frac{\partial f_i}{\partial x_1}(a,b), \quad \frac{\partial f_i}{\partial x_2} = \frac{\partial f_i}{\partial x_2}(a,b).$$

If we define

$$a_{ij} = \frac{\partial f_i}{\partial x_j}(a,b)$$

then the system (4.1) near (a,b) has the form

$$\frac{dX_i}{dt} = a_{i1}X_1 + a_{i2}X_2 + \text{small terms} \quad (i=1,2)$$

when X_1, X_2 are near 0. Hence the trajectories of (4.1) near (a,b) are expected to behave approximately like the trajectories of

$$\frac{dX_i}{dt} = a_{i1}X_1 + a_{i2}X_2, \quad i=1,2. \tag{4.2}$$

Accordingly, the equilibrium point (a,b) of (4.1) is said to be **asymptotically stable** (or, briefly, **stable**) if the equilibrium point $\mathbf{x}=\mathbf{0}$ of (4.2) is asymptotically stable, that is, if the real parts of eigenvalues of the matrix $A = (a_{ij})$ are negative. An equilibrium point which is not stable is called **unstable**.

We conclude that the equilibrium point (a,b) of the system (4.1) is stable if and only if the following inequalities hold at (a,b):

$$\frac{\partial f_1}{\partial x_1} + \frac{\partial f_2}{\partial x_2} < 0,$$

$$\frac{\partial f_1}{\partial x_1}\frac{\partial f_2}{\partial x_2} - \frac{\partial f_1}{\partial x_2}\frac{\partial f_2}{\partial x_1} > 0, \tag{4.3}$$

i.e., trace of $\left(\frac{\partial f_i}{\partial x_j}\right) < 0$ and determinant of $\left(\frac{\partial f_i}{\partial x_j}\right) > 0$. The matrix $\left(\frac{\partial f_i}{\partial x_j}(a,b)\right)$ is called the **Jacobian matrix** at the equilibrium point (a,b).

Example 4.1. Consider the system

$$\frac{dx}{dt} = 2x^2 - xy - 1,$$

$$\frac{dy}{dt} = y - x.$$

We wish to find the equilibrium points and determine their stability. The equilibrium points (x,y) satisfy the equations

$$2x^2 - xy - 1 = 0, \quad y - x = 0.$$

Substituting $y = x$ into the first equation, we get

$$x^2 - 1 = 0.$$

Hence $x = \pm 1$ and the equilibrium points are $(1,1)$ and $(-1,-1)$. In order to determine their stability we compute the Jacobian matrix $J(x,y)$ at (x,y):

$$J(x,y) = \begin{pmatrix} 4x-y & -x \\ -1 & 1 \end{pmatrix} = \begin{pmatrix} 3x & -x \\ -1 & 1 \end{pmatrix}.$$

Then

$$J(1,1) = \begin{pmatrix} 3 & -1 \\ -1 & 1 \end{pmatrix}, \quad J(-1,-1) = \begin{pmatrix} -3 & 1 \\ -1 & 1 \end{pmatrix}.$$

We can get an idea of how the trajectories behave near the equilibrium point $(1,1)$ by computing the eigenvalues of $J(1,1)$, that is, the solutions λ of the characteristic equation at the steady point $(1,1)$,

$$\det(J(1,1) - \lambda I) = 0,$$

or

$$\begin{vmatrix} 3-\lambda & -1 \\ -1 & 1-\lambda \end{vmatrix} = (3-\lambda)(1-\lambda) - 1 = \lambda^2 - 4\lambda + 2 = 0.$$

Clearly, $\lambda_{1,2} = 2 \pm \sqrt{2}$. Since both eigenvalues are positive, the trajectories near $(1,1)$ behave as in Fig. 3.1(A) of an unstable node.

Similarly, the eigenvalues of $J(-1,-1)$ are given by

$$\det(J(-1,-1) - \lambda I) = 0,$$

or

$$\begin{vmatrix} -3-\lambda & 1 \\ -1 & 1-\lambda \end{vmatrix} = (\lambda+3)(\lambda-1) + 1 = \lambda^2 + 2\lambda - 2 = 0,$$

so that $\lambda_{1,2} = 1 \pm \sqrt{3}$. Hence, $\lambda_1 > 0$, $\lambda_2 < 0$, and the trajectories near the steady point $(-1,-1)$ behave as in Fig. 3.1(C) of an unstable saddle point.

Example 4.2. Consider the system

$$\frac{dx}{dt} = -2x^2 + xy,$$

$$\frac{dy}{dt} = -y + x - 3.$$

It is useful to rewrite the first equation by factoring the right-hand side,

$$\frac{dx}{dt} = x(-2x + y).$$

It is then easily seen that the steady points are

$$x = 0, \; y = -3 \quad \text{and} \quad x = -3, \; y = -6.$$

We compute

$$J(x,y) = \begin{pmatrix} -4x+y & x \\ 1 & -1 \end{pmatrix},$$

so that

$$J(0,-3) = \begin{pmatrix} -3 & 0 \\ 1 & -1 \end{pmatrix}, \quad \text{and} \quad J(-3,-6) = \begin{pmatrix} 6 & -3 \\ 1 & -1 \end{pmatrix}.$$

The stationary point $(0,-3)$ is stable since $\text{trace}(J(0,-3))$ $= -4 < 0$ and $\det J(0,-3) = 3 > 0$. The stationary point $(-3,-6)$ is unstable since $\text{trace}(J(-3,-6)) = 5 > 0$. The eigenvalues of $J(0,-3)$ are $\lambda_1 = -1, \lambda_2 = -3$, so the trajectories near the stationary point $(-1,-3)$ behave as in the stable node shown in Fig. 3.1(B). The eigenvalues of $J(-3,-6)$ are

$$\lambda_{1,2} = \frac{5}{2} \pm \sqrt{\frac{25}{4}+3}$$

so $(-3,-6)$ is an unstable saddle point as in Fig. 3.1(C).

Example 4.3. Consider the system

$$\frac{dx}{dt} = x(4-x) - y + I, \quad I \geq 0,$$

$$\frac{dy}{dt} = 2x - y.$$

The steady points are

$$x = 1 \pm \sqrt{1+I}, \quad y = 2x,$$

and

$$J(x,y) = \begin{pmatrix} 4-2x & -1 \\ 2 & -1 \end{pmatrix}.$$

A stationary point (x,y) is stable if and only if

$$\text{trace} J(x,y) = 4 - 2x - 1 < 0, \quad \text{or} \quad x > \frac{3}{2},$$

and

$$\det J(x,y) = -(4-2x) + 2 > 0, \quad \text{or} \quad x > 1.$$

Hence the stationary point $(1+\sqrt{1+I}, 2(1+\sqrt{1+I}))$ is stable and the stationary point $(1-\sqrt{1+I}, 2(1-\sqrt{1+I}))$ is unstable. By computing the eigenvalues of the Jacobians we can determine the behavior of the trajectories about the steady points. For example, if $I = 0$, the characteristic equation about the steady point $(2,4)$ is

$$\lambda^2 + \lambda + 2 = 0$$

and the eigenvalues are $\lambda_{1,2} = -\frac{1}{2} \pm \sqrt{\frac{1}{4} - 2}$. Hence the trajectories behave as in Fig. 3.1(E) of a stable spiral.

Problem 4.1. The system

$$\frac{dx}{dt} = x^2 - y^2,$$
$$\frac{dy}{dt} = x(1 - y)$$

has two nonzero equilibrium points $(1,1), (-1,1)$. Find the eigenvalues of the Jacobian matrix for each of these points, and determine the behavior of the trajectories in terms of the classification described in the graphs in Fig. 3.1.

Problem 4.2. Do the same for the system

$$\frac{dx}{dt} = x - xy^2, \quad \frac{dy}{dt} = y + xy^2 + 1$$

with its steady points $(0, -1), (-2, 1)$.

Problem 4.3. Find the equilibrium points of the system

$$\frac{dx}{dt} = x - xy^2, \quad \frac{dy}{dt} = 1 - x^2 + 2xy$$

and determine their stability

Problem 4.4. Do the same for the system

$$\frac{dx}{dt} = x - x^2 - xy, \quad \frac{dy}{dt} = y - xy - 4y^2.$$

4.1 Numerical Simulations

A general system of two first order ordinary differential equations has the form

$$\begin{cases} \frac{dx_1}{dt} = F_1(x_1, x_2, t), \\ \frac{dx_2}{dt} = F_2(x_1, x_2, t). \end{cases} \tag{4.4}$$

Note that the system (4.4) is slightly different from (4.1) by including t in the right-hand side functions.

Suppose (4.4) is a linear system, namely, it can be written in the form:

$$\begin{cases} \frac{dx_1}{dt} = a_{11}(t)x_1 + a_{12}(t)x_2 + b_1(t) \\ \frac{dx_2}{dt} = a_{21}(t)x_1 + a_{22}(t)x_2 + b_2(t), \end{cases}$$

or equivalently,

$$x' = A(t)x + b(t)$$

where

$$x = \begin{pmatrix} x_1(t) \\ x_2(t) \end{pmatrix}, \; b(t) = \begin{pmatrix} b_1(t) \\ b_2(t) \end{pmatrix}, \; A(t) = \begin{pmatrix} a_{11}(t) & a_{12}(t) \\ a_{21}(t) & a_{22}(t) \end{pmatrix}.$$

We can use Algorithms 3.1 and 3.2 to solve the system numerically (the matrix A in Algorithm 3.2 may have to be changed to a function of t if A is not constant).

However, if the system (4.4) is nonlinear, then we can no longer use Algorithm 3.2. In the following example, we show how to change the code when the system is nonlinear.

Example 4.4. Solve the system

$$\frac{dx_1}{dt} = x_1^2 - x_1 x_2 + t,$$

$$\frac{dx_2}{dt} = -x_1 + x_2^2.$$

with $(x_1, x_2) = (1,1)$ and $0 \le t \le 1$. Sample codes are in Algorithms 4.1 and 4.2.

Algorithm 4.1. Main file for solving Example 4.4 (main_NonlinearDiffEqns.m)

```
x_ini = [1,1]'; % the initial condition is a vector
tspan = [0,1] ; % time span
[t,z] = ode45('fun_NonlinearDiffEqns', tspan, x_ini);
% output t is a column vector, and the output z is a matrix
% containing two columns of the same lengths as vector t

plot(t,z)
legend('x_1','x_2')    % add legend at the corner of the figure to
                       % distinguish two curves
```

Algorithm 4.2. fun_NonlinearDiffEqns.m

```
function dx = fun_NonlinearDiffEqns(t,x)
% the input is t (scalar) and x (a 2-by-1 column vector)

dx = zeros(2,1);
% the output dx is preset to be a 2-by-1 zero column vector
dx(1) = x(1)^2 - x(1)*x(2) + t;  % the first component of dx
dx(2) = -x(1) + x(2)^2;          % the second component of dx
```

Problem 4.5. Solve numerically the system

$$\frac{dx}{dt} = xy - 2y,$$
$$\frac{dy}{dt} = xy + x,$$

with $x(0) = 1$, $y(0) = 1$, for $0 \leq t \leq 3$.

Problem 4.6. Solve numerically the system

$$\frac{dx}{dt} = x - xy^2 + t,$$
$$\frac{dy}{dt} = y + xy,$$

with $x(0) = 1$, $y(0) = 1$, for $0 \leq t \leq 3$.

Problem 4.7. Solve numerically the system

$$\frac{dx}{dt} = -xy,$$
$$\frac{dy}{dt} = (1 - x)(1 + y),$$

with $x(0) = 2$, $y(0) = 0$, for $0 \leq t \leq 4$.

In this chapter, we have learned that we can use the eigenvalues of the Jacobian to determine whether a stationary point is stable or not. In MATLAB, we can easily evaluate the eigenvalues and eigenvectors.

Example 4.5. Given the matrix

$$A = \begin{pmatrix} 1 & 1 \\ 4 & 1 \end{pmatrix}$$

we can use the following commands to obtain the eigenvalues and eigenvectors.
```
>> A=[1 1;4 1];
>> [V,D]=eig(A)
V =
   0.4472 -0.4472
   0.8944 0.8944
D =
   3.0000 0
   0        -1.0000
```
The output V contains two column vectors, the eigenvectors of A, and their corresponding eigenvalues are in the diagonal of matrix D.

Therefore, if we are given the following system

$$\frac{d\mathbf{x}}{dt} = \begin{pmatrix} 1 & 1 \\ 4 & 1 \end{pmatrix} \mathbf{x} \tag{4.5}$$

we can easily construct the exact solution

$$\mathbf{x} = c_1 \begin{pmatrix} 1 \\ 2 \end{pmatrix} e^{3t} + c_2 \begin{pmatrix} 1 \\ -2 \end{pmatrix} e^{-t}.$$

The origin is a saddle point and it is unstable. Figure 4.1 is an illustration of such an unstable saddle point.

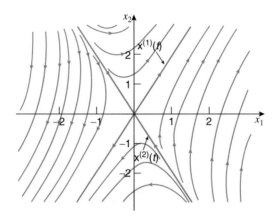

Fig. 4.1: Unstable saddle point for the system (4.5).

Problem 4.8. Consider the system

$$\frac{d\mathbf{x}}{dt} = \begin{pmatrix} -3 & \sqrt{2} \\ \sqrt{2} & -2 \end{pmatrix} \mathbf{x}.$$

Obtain the exact solution by hand or by using the commands introduced above. Is the origin (the stationary point) stable?

Problem 4.9. Solve the system

$$\frac{d\mathbf{x}}{dt} = \begin{pmatrix} -\frac{1}{2} & 1 \\ -1 & -\frac{1}{2} \end{pmatrix} \mathbf{x}.$$

Is the origin (the stationary point) stable?

Problem 4.10. Solve the system

$$\frac{d\mathbf{x}}{dt} = \begin{pmatrix} 1 & -1 \\ 1 & 3 \end{pmatrix} \mathbf{x}.$$

Is the origin (the stationary point) stable?

Chapter 5
Predator–Prey Models

A predator is an organism that eats another organism. A prey is an organism that a predator eats. In ecology, a **predation** is a biological interaction where a predator feeds on a prey. Predation occurs in a wide variety of scenarios, for instance in wild life interactions (lions hunting zebras, foxes hunting rabbits), in herbivore–plant interactions (cows grazing), and in parasite–host interactions.

If the predator is to survive over many generations, it must ensure that it consumes sufficient amount of prey, otherwise its population will decrease over time and will eventually disappear. On the other hand, if the predator over-consumes the prey, the prey population will decrease and disappear, and then the predator will also die out, from starvation.

Thus the question arises: what is the best strategy of the predator that will ensure its survival. This question is very important to ecologists who are concerned with biodiversity. But it is also an important question in the food industry; for example, in the context of fishing, with humans as predator and fish as prey, what is the sustainable amount of fish harvesting?

In this chapter we use mathematics to provide answers to these questions. We begin with a simple example of predator–prey interaction.

We denote by x the density of a prey, that is, the number of prey animals per unit area on land (or volume, in sea) and by y the density of predators. We denote by a the net growth rate in x (birth minus natural death), and by c the net death rate of predators. The growth of predators is assumed to depend only on its consumption of the prey as food. Predation occurs when predator comes into close contact with prey, and we take this encounter to occur at an average rate b. Hence

$$\frac{dx}{dt} = ax - bxy. \tag{5.1}$$

The growth of predators is proportional to bxy (say, by a factor of d/b), so that

Electronic supplementary material The online version of this chapter (doi: 10.1007/978-3-319-29638-8_5) contains supplementary material, which is available to authorized users.

C.-S. Chou, A. Friedman, *Introduction to Mathematical Biology*,
Springer Undergraduate Texts in Mathematics and Technology,
DOI 10.1007/978-3-319-29638-8_5

$$\frac{dy}{dt} = dxy - cy. \tag{5.2}$$

In terms of dimensions,

$$[a] = \frac{1}{\text{time}}, \quad [b] = \frac{1}{\text{density of predator}}\frac{1}{\text{time}},$$

and

$$[c] = \frac{1}{\text{time}}, \quad [d] = \frac{1}{\text{density of prey}}\frac{1}{\text{time}}.$$

The system (5.1)–(5.2) has two equilibrium points. The first one is $(0,0)$; this corresponds to a situation where both species die. This equilibrium point is unstable. Indeed the Jacobian matrix at $(0,0)$ is

$$\begin{pmatrix} a & 0 \\ 0 & -c \end{pmatrix}$$

and one of the eigenvalues, namely a, is positive.

The second equilibrium point is $(\frac{c}{d}, \frac{a}{b})$ and the Jacobian matrix at this point is

$$\begin{pmatrix} 0 & \frac{-bc}{d} \\ \frac{ad}{b} & 0 \end{pmatrix}.$$

The corresponding eigenvalues are $\lambda = \pm i\sqrt{ac}$. According to Fig. 3.1(F) the portrait of all trajectories are circles. We conclude: The predator and prey will both survive forever, and their population will undergo periodic (seasonal) oscillations.

The system (5.1)–(5.2) is an example of what is known as **Lotka-Volterra** equations. One can introduce various variants into these equations. For example, if the prey population is quite congested, we may want to use the logistic growth for the prey (recall that logistic growth is introduced in Eq. (2.17)).

More general models of predator–prey are written in the form

$$\frac{dx}{dt} = xf(x,y), \quad \frac{dy}{dt} = yg(x,y),$$

where x is the prey and y is the predator, $\partial f/\partial y < 0, \partial g/\partial x > 0$, and $\partial f/\partial x < 0$ for large x, $\partial g/\partial y < 0$ for large y. The first two inequalities mean that the prey population is depleted by the predator and the predator population is increased by feeding on the prey. The last two inequalities represent natural death due to the logistic growth model.

Example 5.1. Consider the predator–prey system

$$\frac{dx}{dt} = ax(1 - \frac{x}{A}) - bxy,$$ (5.3)

$$\frac{dy}{dt} = dxy(1 - \frac{y}{B}) - cy,$$ (5.4)

where A and B are the carrying capacities for the prey x and the predator y, respectively. In order to compute the steady points and determine their stability we conveniently factor out x in (5.3) and y in (5.4), rewriting these equations in the form

$$\frac{dx}{dt} = x[a(1 - \frac{x}{A}) - by],$$ (5.5)

$$\frac{dy}{dt} = y[dx(1 - \frac{y}{B}) - c].$$ (5.6)

Clearly, $(x,y) = (0,0)$ is a steady point with the Jacobian

$$\begin{pmatrix} a & 0 \\ 0 & -c \end{pmatrix},$$

so $(0,0)$ is unstable. The point $(\bar{x},\bar{y}) = (A,0)$ is another steady point with Jacobian

$$\begin{pmatrix} -a & -Ab \\ 0 & dA - c \end{pmatrix}.$$

Hence the steady state $(A,0)$, where only the prey survives, is stable if $dA - c < 0$.

The nonzero steady point $(\bar{x},\bar{y})$ (where $\bar{x} > 0$ and $\bar{y} > 0$) are determined by solving the equations

$$a(1 - \frac{\bar{x}}{A}) - b\bar{y} = 0,$$ (5.7)

$$d\bar{x}(1 - \frac{\bar{y}}{B}) - c = 0.$$ (5.8)

But before computing these points let us compute the Jacobian at $(\bar{x},\bar{y})$. In view of (5.7) we have

$$\frac{\partial}{\partial x}\{x[a(1 - \frac{x}{A}) - by]\}|_{(\bar{x},\bar{y})} = \{x\frac{\partial}{\partial x}[a(1 - \frac{x}{A}) - by]\}|_{(\bar{x},\bar{y})} = -\bar{x}\frac{a}{A};$$

similarly

$$\frac{\partial}{\partial y}\{y[dx(1 - \frac{y}{B}) - c]\}|_{(\bar{x},\bar{y})} = \{y\frac{\partial}{\partial y}[dx(1 - \frac{y}{B}) - c]\}|_{(\bar{x},\bar{y})} = -\bar{y}\frac{d\bar{x}}{B}.$$

where we have used (5.8). Hence

$$J(\bar{x}, \bar{y}) = \begin{pmatrix} -\frac{\bar{x}a}{A} & -b\bar{x} \\ \bar{y}d(1 - \frac{\bar{y}}{B}) & -\frac{\bar{y}d\bar{x}}{B} \end{pmatrix}. \tag{5.9}$$

We immediately see that $\text{trace}(J(\bar{x}, \bar{y})) < 0$. Hence $(\bar{x}, \bar{y})$ is stable if and only if $\det J(\bar{x}, y) > 0$, where

$$\det J(\bar{x}, \bar{y}) = d\bar{x}\bar{y}[\frac{a\bar{x}}{AB} + b(1 - \frac{\bar{y}}{B})]. \tag{5.10}$$

To search for other steady points we substitute

$$\bar{y} = \frac{a}{b}(1 - \frac{\bar{x}}{A}) \tag{5.11}$$

from (5.7) into (5.8) and obtain a quadratic equation for $\bar{x}$:

$$\alpha\bar{x}^2 + \beta\bar{x} - c = 0,$$

where

$$\alpha = \frac{ad}{bAB}, \quad \beta = d(1 - \frac{a}{bB}).$$

The only positive solution is

$$\bar{x} = \frac{1}{2\alpha}[-\beta + \sqrt{\beta^2 + 4ac}]. \tag{5.12}$$

Since A is the carrying capacity of x, it is biologically natural to assume that $\bar{x} < A$. Actually, if $x(0) < A$ and $y(t)$ is assumed to be positive for all $t > 0$, then $x(t)$ will remain less than A for all $t > 0$. We can show this by contradiction: otherwise there is a first time, t_0, when $x(t)$ becomes equal to A, so that $x(t_0) = A$ and $\frac{dx}{dt}(t_0) \geq 0$. But, by (5.5)

$$\frac{dx}{dt}(t_0) = -bx(t_0)y(t_0) < 0,$$

which is a contradiction.

Similarly, it is natural to assume that $\bar{y} < B$. Hence $(\bar{x}, \bar{y})$ is a biologically relevant steady state with $\bar{y} > 0$ if and only if $\bar{y}$ is given by (5.11), and the inequalities

$$\bar{x} < A, \quad \frac{a}{b}(1 - \frac{\bar{x}}{A}) < B \tag{5.13}$$

hold, where $\bar{x}$ is defined by (5.12). These inequalities hold, for instance, if $\frac{a}{b} < B$ and c is small.

The above example is instructive in two ways. First, it shows that sometimes it is better to compute the Jacobian at $(\bar{x}, \bar{y})$ before actually computing the steady point $(\bar{x}, \bar{y})$ whose expression could be complicated. Second, it shows that by factoring out *x* or *y* in the differential equations we are able to compute the Jacobian more easily. This last remark will be very useful in future computations, so we shall refer it as the 'factorization rule' and formulate it for general systems of equations.

Factorization Rule

Consider a system (4.1) where the f_i can be factored as follows:

$$f_1(x_1,x_2) = x_1 g_1(x_1,x_2), \quad f_2(x_1,x_2) = x_2 g_2(x_1,x_2),$$

so that

$$\frac{dx_1}{dt} = x_1 g_1(x_1,x_2), \quad \frac{dx_2}{dt} = x_2 g_2(x_1,x_2)$$

In this case there are equilibrium points $P_1 = (0,0), P_2 = (0,\tilde{x}_2)$ if $g_2(0,\tilde{x}_2) = 0$, $P_3 = (\tilde{x}_1,0)$ if $g_1(\tilde{x}_1,0) = 0$, and $P_4(\tilde{x}_1,\tilde{x}_2)$ if $g_1(\tilde{x}_1,\tilde{x}_2) = 0$, $g_2(\tilde{x}_1,\tilde{x}_2) = 0$. We can then quickly compute the Jacobian matrix $J(P_i)$ at each point P_i. For example, to compute $J(P_4)$ when $\tilde{x}_1 > 0, \tilde{x}_2 > 0$, we notice that since $g_1 = g_2 = 0$ at P_4,

$$J(P_4) = \begin{pmatrix} x_1 \frac{\partial g_1}{\partial x_1} & x_1 \frac{\partial g_1}{\partial x_2} \\ x_2 \frac{\partial g_2}{\partial x_1} & x_2 \frac{\partial g_1}{\partial x_2} \end{pmatrix}_{(\tilde{x}_1,\tilde{x}_2)}.$$

Similarly,

$$J(P_1) = \begin{pmatrix} g_1(0,0) & 0 \\ 0 & g_2(0,0) \end{pmatrix},$$

$$J(P_2) = \begin{pmatrix} g_1 & 0 \\ x_2 \frac{\partial g_2}{\partial x_1} & x_2 \frac{\partial g_1}{\partial x_2} \end{pmatrix}_{(0,\tilde{x}_2)} \quad \text{when } \tilde{x}_2 > 0,$$

and

$$J(P_3) = \begin{pmatrix} x_1 \frac{\partial g_1}{\partial x_1} & x_1 \frac{\partial g_1}{\partial x_2} \\ 0 & g_2 \end{pmatrix}_{(\tilde{x}_1,0)} \quad \text{when } \tilde{x}_1 > 0.$$

Example 5.2. Plant–herbivore model. The herbivore H feeds on plant P, which grows at rate r. We take the consumption rate of the plant to be

$$\frac{\sigma P}{1+P} H;$$

this means that, at small amount of P, H consumes P at a linear rate σP, but the rate of consumption by H is limited and, for simplicity, we assume that it cannot exceed σ. Thus,

$$\frac{dP}{dt} = rP - \sigma \frac{P}{1+P} H. \tag{5.14}$$

The equation for the herbivore is

$$\frac{dH}{dt} = \lambda \sigma \frac{P}{1+P} H - dH. \tag{5.15}$$

Here d is the death rate of H, and λ is the **yield constant**, that is,

$$\lambda = \frac{\text{mass of herbivore formed}}{\text{mass of plant used}};$$

naturally $\lambda < 1$. Note that if $\lambda \sigma < d$, that is, if the growth rate by consumption is less than the death rate, then $\frac{dH}{dt} < 0$ and the herbivore will die out. We rewrite the system (5.14)–(5.15) in the more convenient from

$$\frac{dP}{dt} = P(r - \frac{\sigma}{1+P} H), \tag{5.16}$$

$$\frac{dH}{dt} = H(\lambda \frac{\sigma P}{1+P} - d), \tag{5.17}$$

and assume that $\lambda \sigma > d$. Then the steady points are $(0,0)$ and $(\bar{P}, \bar{H})$, where

$$\bar{P} = \frac{d}{\lambda \sigma - d}, \quad \bar{H} = \frac{\lambda r}{\lambda \sigma - d}.$$

Since

$$J(0,0) = \begin{pmatrix} r & 0 \\ 0 & -d \end{pmatrix},$$

the steady point $(0,0)$ is unstable. Using the factorization rule we find that

$$J(\bar{P}, \bar{H}) = \begin{pmatrix} \frac{\bar{P}\sigma\bar{H}}{(1+\bar{P})^2} & -\frac{\sigma\bar{P}}{1+\bar{P}} \\ \frac{\lambda\sigma}{(1+\bar{P})^2} & 0 \end{pmatrix}.$$

Since trace $J(\bar{P}, \bar{H}) > 0$ and det $J(\bar{P}, \bar{H}) > 0$, and both eigenvalues of the characteristic equation have negative real parts, so $(\bar{P}, \bar{H})$ is an unstable node. We conclude that the plant–herbivore model (5.14)–(5.15) has no stable steady states. In order to understand this situation better we look at the dynamics of system (5.16)–(5.17) on the P-H phase plane. The solutions of the system form trajectories on the phase plane, as depicted in Fig. 3.1; here we will analyze the direction of the trajectories, that is, $(\frac{dP}{dt}, \frac{dH}{dt})$ on the phase plane in order to get an idea of how the trajectories themselves look like. We introduce the nullclines (see definition in Chapter 4)

$$\Gamma_1 : r - \frac{\sigma}{1+P} H = 0, \quad \text{where } \frac{dP}{dt} = 0 \text{ on } \Gamma_1$$

$$\Gamma_2 : \frac{\lambda \sigma P}{1+P} - d = 0, \quad \text{where } \frac{dH}{dt} = 0 \text{ on } \Gamma_2.$$

The arrows in Figure 5.1 show the direction of the trajectories. Notice that

$$\frac{dH}{dt} > 0 \quad \text{if} \quad \lambda\sigma\frac{P}{1+P} - d > 0,$$

i.e., if $(\lambda\sigma - d)P - d > 0$, or

$$P > \frac{d}{\lambda\sigma - d}.$$

So on Γ_1

$$\frac{dP}{dt} = 0, \quad \text{and} \quad \frac{dH}{dt} > 0 \text{ if } P > \frac{d}{\lambda\sigma - d};$$

consequently the vector

$$\left(\frac{dP}{dt}, \frac{dH}{dt}\right)$$

points vertically upward at points of Γ_1 where $P > \frac{d}{\lambda\sigma - d}$. Similarly, on Γ_1,

$$\frac{dP}{dt} = 0, \quad \text{and} \quad \frac{dH}{dt} < 0 \text{ if } P < \frac{d}{\lambda\sigma - d},$$

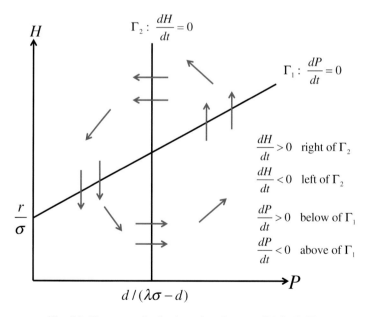

Fig. 5.1: Phase portrait of trajectories of system (5.16)–(5.17).

so that the vector $(dP/dt, dH/dt)$ points vertically downward. In the same way we can see that on Γ_2, where $P = \frac{d}{\lambda\sigma - d}$ and $\frac{dH}{dt} = 0$, the vector $(dP/dt, dH/dt)$ points horizontally, to the right below Γ_1 (where $\frac{dP}{dt} > 0$) and to the left above Γ_1 (where $\frac{dP}{dt} < 0$). Next, in the region below Γ_1 and to the right of Γ_2,

$$\frac{dP}{dt} > 0, \quad \frac{dH}{dt} > 0$$

so that the vector $(dP/dt, dH/dt)$ points upward and to the right, as shown in Fig. 5.1. Similarly, the direction of the vector is upward to the left in the region above Γ_1 and to the right of Γ_2. On the left of Γ_2, the vector $(dP/dt, dH/dt)$ points downward: either to the left (above Γ_1) or to the right (below Γ_1). The arrows in Fig. 5.1 schematically summarize the above considerations.

We see that the phase portrait of the nonlinear system (5.16)–(5.17) is similar to the phase portrait of an unstable spiral, as in Fig. 3.1(D).

Problem 5.1. Consider a predator–prey model where the carrying capacity of the predator y depends linearly on the density of the prey:

$$\frac{dx}{dt} = ax(1 - \frac{x}{A}) - bxy,$$

$$\frac{dy}{dt} = dy(1 - \frac{y}{1+x}).$$

Find the steady points and determine their stability.

The **Allee** effect refers to the biological fact that increased fitness correlates positively with higher (but not overcrowded) population, or that 'undercrowding' decreases fitness. More specifically, if the size of a population is below a threshold then it is destined for extinction. Endangered species are often subject to the Allee effect.

Consider a predator–prey model where the prey is subject to the Allee effect,

$$\frac{dx}{dt} = rx(x - \alpha)(1 - x) - \sigma xy, \quad (0 < \alpha < 1), \tag{5.18}$$

that is, if the population $x(t)$ decreases below the threshold $x = \alpha$, then $x(t)$ will decrease to zero as $t \to \infty$. The predator y satisfies the equation

$$\frac{dy}{dt} = \lambda \sigma xy - \delta y, \tag{5.19}$$

where λ is a constant.

Problem 5.2. The point $(0,0)$ is an equilibrium point of the system (5.18)–(5.19). Determine whether it is asymptotically stable.

Problem 5.3. Show that if $\alpha < \frac{\delta}{\lambda\sigma} < 1$, then the system (5.18)–(5.19) has a second equilibrium point $(\bar{x}, \bar{y}) = (\frac{\delta}{\lambda\sigma}, \frac{r}{\sigma}(\frac{\delta}{\lambda\sigma} - \alpha)(1 - \frac{\delta}{\lambda\sigma}))$, and it is stable if

$$\frac{\delta}{\lambda\sigma} > \frac{1+\alpha}{2}.$$

This result shows that for the predator to survive, the prey must be allowed to survive, and the predator must adjust its maximum eating rate, σ, so that

$$\frac{\delta}{\lambda} < \sigma < \frac{\delta}{\lambda}\frac{2}{1+\alpha}.$$

If the Allee threshold, α, deteriorates and approaches 1, the predator must then decrease its rate of consumption of the prey and bring it closer to δ/λ, otherwise it will become extinct.

5.1 Numerical Simulations

The following algorithms 5.1 and 5.2 simulate (5.1)–(5.2). These codes demonstrate how to implement nonlinear systems (also see Chapter 4). In these codes, there are several parameters (a,b,c,d) which may be changed from simulation to simulation. We here define them as **global** variables, which can be recognized in files declaring them as global variables. It is convenient to use the global variables to define parameters that we would like to tune in models: we only have to assign values in the main file, without changing their numbers in the function files. But we also need to be careful with the names of these global parameters to prevent changing them accidentally in the code or using the same names to define other variables.

Algorithm 5.1. Main file for model (5.1)–(5.2) (main_predator_prey.m)

```
%%% This code simulates model (5.1)-(5.2).
close all, % close all the figure windows
clear all, % clear all the variables

%% define global variables
global a b c d
%% starting and final time
t0 = 0; tfinal = 5;
%% paramters
a = 5; b = 2; c = 9; d = 1;
%% initial conditions
v0 = [10,5];
[t,v] = ode45('fun_predator_prey',[t0,tfinal],v0);
subplot(2,1,1)
plot(t,v(:,1)) % plot the evolution of x
xlabel t, ylabel x
subplot(2,1,2)
plot(t,v(:,2)) % plot the evolution of y
xlabel t, ylabel y
```

Problem 5.4. Plot the time evolution of model (5.1)–(5.2) with $a = 5, b = 2$, $c = 9, d = 1$ starting from $(10,5)$, for time from 0 to 5.

Algorithm 5.2. fun_predator_prey.m

```
% This is the function file called by main_predator_prey.m
function dy = fun_predator_prey(t,v)
%% define global variables
global a b c d
dy = zeros(2,1);
dy(1)  = a*v(1)  - b*v(1)*v(2);
dy(2)  = -c*v(2) + d*v(1)*v(2);
```

Problem 5.5. Hand draw the phase portrait for (5.1)–(5.2) with $a = 5, b = 2$, $c = 9, d = 1$ starting from several points near the nonzero steady point.

Problem 5.6. Change the codes (adding global variables A and B in both files, define the values in the main file, and change dy(1) and dy(2) in fun_predator_prey.m) to implement (5.3)–(5.4). Plot the time evolution with $a = 5, b = 2, c = 1, d = 1$, $A = 2, B = 3$ starting from $(10, 5)$, for time from 0 to 10. What is the steady state you see from the simulation (you can print out the last row of the solution vector to get x and y). Verify the stability condition using this set of parameters.

5.1.1 Revisiting Euler Method for Solving ODE – Consistency and Convergence

We introduce some basic concepts in numerical analysis. These concepts will be important in general for choosing an appropriate scheme to use and assess the error of the selected algorithm. We will practice to write our own time integrator to solve ODE instead of using ode45 in MATLAB.

Consider a differential equation

$$\frac{dx}{dt} = f(x,t), \quad t \ge t_0, \quad x(t_0) = x_0, \tag{5.20}$$

where f is a continuously differentiable function in x and t and x_0 is an initial condition. Note that although here we consider a single equation where x is a real-valued function, the following discussion can be easily generalized to systems in which x and f represent vector-valued functions. There are various ways to derive Euler method; here we give one derivation based on interpolation.

Recall that forward Euler method for solving (5.20) has the formula (see Eq. 2.24)

$$X_{n+1} = X_n + hf(X_n, t_n), \tag{5.21}$$

where X_n denotes the approximate solution at time t_n, and $t_0 < t_1 < \cdots < t_N = T$ are equi-distanced grid points with $h = t_{n+1} - t_n$. These types of schemes are called **explicit schemes** because the solution X_{n+1} is explicitly defined as a function of X_n. In other words, knowing X_n, one can explicitly compute X_{n+1}. Furthermore, it is called a **single step** method because it requires only solution at one time step in order to compute the solution at the following time step.

In order to understand how good the numerical solution is, we define **local truncation error**, d_n, to measure how closely the difference operator approximates the differential operator, for forward Euler method:

$$d_n \equiv \frac{x(t_{n+1}) - x(t_n)}{h} - f(x(t_n), t_n) = \frac{h}{2} x''(\bar{t}_n) + O(h^2),$$

where $\bar{t}_n$ is some point in the interval $[t_n, t_{n+1}]$. In other words, the truncation error is the measure of error by plugging in the exact solution into our numerical scheme. The truncation error analysis can be easily obtained by using Taylor expansion around $t = t_n$. If a method has the local truncation error $O(h^p)$, we say that the method is p-th order accurate. The forward Euler method is first order accurate because the leading term of d_n is of order h.

However, the real goal is not consistency but **convergence**. Assume Nh is bounded independently of N. The method is said to be **convergent of order** p if the **global** error e_n, where $e_n = X_n - x(t_n)$, $e_0 = 0$, satisfies

$$e_n = O(h^p), \quad n = 1, 2, \cdots, N.$$

That is, we hope that, after the accumulation of the local errors through all the steps, the errors can still be controlled and bounded by $O(h^p)$.

Example 5.3. Consider the problem

$$\frac{dy}{dt} = \lambda y, \quad y(0) = y_0.$$

We know that the exact solution is $y(t) = y_0 e^{\lambda t}$. If $\lambda < 0$, we expect that $|y(t)|$ exponentially decreases to 0. Let's apply forward Euler method to this problem, which we call the 'test problem.' We get

$$X_{n+1} = X_n + h\lambda X_n, \quad n = 0, 1, \ldots \tag{5.22}$$

with $X_0 = y_0$ and h being the time step in our discretization. Simplifying (5.22), we obtain

$$X_{n+1} = X_n(1 + h\lambda), \quad n = 0, 1, \ldots$$

and therefore

$$X_n = (1 + h\lambda)^n X_0 = (1 + h\lambda)^n y_0, \quad n = 0, 1, \ldots$$

Recall that we expect $|y(t)|$ to decrease exponentially, so we require the approximation to satisfy $|X_{n+1}| < |X_n|$, that is, $|1 + h\lambda| < 1$ ($-1 < 1 + h\lambda < 1$). So in order to obtain the desired behavior of the solution, we need to require that

$$h < \frac{-2}{\lambda} \quad (\lambda < 0). \tag{5.23}$$

The condition (5.23) is a condition imposed on the time step, which we call the **stability condition**. If this condition is violated, then our numerical solution blows up, as can be seen in the numerical experiment in Problem 5.7.

Problem 5.7. Consider the test problem with $\lambda = 20$ and $y_0 = 1$. The sample MATLAB codes can be found in Algorithm 5.3. (a) Derive the stability condition. (b) Test $h = 0.01, 0.05, 0.1, 0.2$, with final time $T = 1$. Describe what you see and explain it theoretically.

Algorithm 5.3. Forward_Euler method for the test problem (forward_Euler.m)

```
% This is a code to solve dy/dt = lambda*(y), 0<t<1, y(t=0)=1
% (lambda=20)   using forward Euler

clear all;
lambda = -20;
h = .0005;
t = 0:h:1;
Nt = length(t);
y = zeros(Nt,2); % preset y as a zero matrix with the same length
                 % as t
y(1) = 1;             % initial condition; index starts from 1

for i = 1:(Nt-1)
    y(i+1) = y(i)+h*(lambda*(y(i)));
end

error = abs(y(end)-exp(lambda*t(end)))
plot(t,y), hold on
plot(t,exp(lambda*t),'r')
```

Problem 5.8. Consider the scalar problem

$$y' = -5ty^2 + \frac{5}{t} - \frac{1}{t^2}, \quad y(1) = 1.$$

(a) Verify that $y(t) = \frac{1}{t}$ is a solution to the problem. (b) Use forward Euler method until $t = 10$ (modify forward_Euler.m). Define the error to be the absolute value of the difference between the exact solution (in this problem, the exact solution is $\frac{1}{10}$, for $t = 10$) and the numerical solution (in the MATLAB code it will be 'y(end)'). Compute the error at $t = 10$ using $h = 0.0025, 0.005, 0.01, 0.02$. Verify that this method is first order accurate based on the errors (the error should decrease by half when you decrease h by half).

As mentioned above, the method (5.21) can be applied to systems, where x and f are vectors. Algorithm 5.4 is a sample code using forward Euler method to simulate (5.1)–(5.2). Note that in the code, we used the function 'fun_predator_pray.m' that was defined in Algorithm 5.2. Since forward Euler is first order, we can see that when we increase h, the solution may be less accurate, and the shape is not as expected. But this inaccuracy will be gone once the time step is small enough.

Algorithm 5.4. Forward Euler method for predator–pray model (5.1)–(5.2) (forward_Euler_predator_prey.m)

```
% This code simulates model (5.1)-(5.2) using Forward Euler method.
close all, % close all the figure windows
clear all, % clear all the variables
%% define global parameters
global a b c d

%% starting and final time
t0 = 0; tfinal = 5;

%% paramters
a = 5; b = 2; c = 9; d = 1;

%% set up time step and vectors
h = .001;               % time step
t = t0 : h : tfinal; % discrete time steps
Nt = length(t);         % total number of time steps
v = zeros(Nt,2); % preset v as a zero matrix
% Note that v has two columns, each representing one variable

%% initial conditions
v(1,:) = [10,5];

for i = 1:(Nt-1)
    time = t(i); % current time
        z = v(i,:); % approximate solution at the current time step
      rhs = fun_predator_prey(time,z)';
   v(i+1,:) = z + h*rhs;
end

%% plot
subplot(2,1,1)
plot(t,v(:,1)) % plot the evolution of x
xlabel t, ylabel x
subplot(2,1,2)
plot(t,v(:,2)) % plot the evolution of y
xlabel t, ylabel y
```

Chapter 6
Two Competing Populations

Competition is an interaction between organisms, or species, sharing resources that are in limited supply. This is an important topic in ecology. The 'competitive exclusion principle' asserts that species less suited to compete will either adapt or die out. In aggressive competition one species may attempt to kill the other. This situation occurs, for example, among some species of ants, and some species or yeast. When enough data is known about the history of a specific competition between two species, mathematics can then be used to predict whether both species will survive and coexist or whether one of them will die out.

In this chapter we consider some examples of competing populations and determine, using mathematics, whether one or both species will survive. We begin with the following model:

$$\frac{dx}{dt} = r_1 x(1 - \frac{x}{k_1}) - b_1 xy, \tag{6.1}$$

$$\frac{dy}{dt} = r_2 y(1 - \frac{y}{k_2}) - b_2 xy. \tag{6.2}$$

In Eq. (6.1), r_1 is the growth rate of species x, k_1 is the carrying capacity which limits its growth, and b_1 is the rate by which the competitor y kills x. Eq. (6.2) has a similar interpretation.

The system (6.1)–(6.2) has equilibrium points

$$(0,0), \quad (k_1,0), \quad (0,k_2). \tag{6.3}$$

Note that the equilibrium point $(k_1,0)$ means that the second population becomes extinct. Similarly, $(0,k_2)$ corresponds to a situation where the first population becomes extinct.

Electronic supplementary material The online version of this chapter (doi: 10.1007/ 978-3-319-29638-8_6) contains supplementary material, which is available to authorized users.

© Springer International Publishing Switzerland 2016
C.-S. Chou, A. Friedman, *Introduction to Mathematical Biology*,
Springer Undergraduate Texts in Mathematics and Technology,
DOI 10.1007/978-3-319-29638-8_6

In order to determine whether there exist additional equilibrium points, we must solve the equations

$$r_1\left(1 - \frac{x}{k_1}\right) - b_1 y = 0,$$

$$r_2\left(1 - \frac{y}{k_2}\right) - b_2 x = 0.$$

We rewrite the system in the form

$$\frac{r_1}{k_1}x + b_1 y = r_1,$$

$$b_2 x + \frac{r_2}{k_2}y = r_2.$$

The determinant D of the system is

$$D = \begin{vmatrix} \frac{r_1}{k_1} & b_1 \\ b_2 & \frac{r_2}{k_2} \end{vmatrix} = \frac{r_1 r_2}{k_1 k_2} - b_1 b_2 = \frac{r_1 r_2}{k_1 k_2}(1 - \beta_1 \beta_2),$$

where $\beta_1 = \frac{k_1 b_1}{r_1}$, $\beta_2 = \frac{k_2 b_2}{r_2}$, and the solution of the system is given by

$$x = \frac{1}{D}\begin{vmatrix} r_1 & b_1 \\ r_2 & \frac{r_2}{k_2} \end{vmatrix} = \frac{1}{D}\left(\frac{r_1 r_2}{k_2} - r_2 b_1\right) = \frac{1}{1 - \beta_1 \beta_2}\frac{k_1 k_2}{r_1 r_2}\left(\frac{r_1 r_2}{k_2} - r_2 b_1\right) = \frac{k_1 - k_2 \beta_1}{1 - \beta_1 \beta_2},$$

and, similarly,

$$y = \frac{k_2 - k_1 \beta_2}{1 - \beta_1 \beta_2}.$$

Thus the steady point at which both populations may exist is given by

$$\left(\frac{\beta_1 k_2 - k_1}{\beta_1 \beta_2 - 1}, \frac{\beta_2 k_1 - k_2}{\beta_1 \beta_2 - 1}\right), \quad \text{where } \beta_i = \frac{k_i b_i}{r_i} \ (i = 1, 2). \tag{6.4}$$

This steady point is of biological relevance only if the two components are positive, which occurs only when either

$$k_1 > \frac{r_2}{b_2}, \quad k_2 > \frac{r_1}{b_1}$$

or

$$k_1 < \frac{r_2}{b_2}, \quad k_2 < \frac{r_1}{b_1}.$$

Problem 6.1. Determine whether the equilibrium points in (6.3) are stable.

Problem 6.2. Show that the steady point defined in (6.4) is stable if $k_1 < \frac{r_2}{b_2}$ and $k_2 < \frac{r_1}{b_1}$.

This result means that both species will coexist provided that the rate of killing, b_j, is less than r_i/k_j, the rate of growth divided by the carrying capacity, for $j = 1, i = 2$ and for $j = 2, i = 1$.

In the next example two species are competing for space. Consider for example grass (x) and weed (y) growing in the same field. They share some resources, e.g., nutrients from the ground. But they also receive resources independently from each other, e.g., sunshine and rain. Thus they only partially infringe upon each other in terms of the medium carrying capacity which supports their growth. We can model their dynamics as follows:

$$\frac{dx}{dt} = r_1 x (1 - \frac{x + \alpha y}{K}) - \mu_1 x, \tag{6.5}$$

$$\frac{dy}{dt} = r_2 y (1 - \frac{\beta x + y}{K}) - \mu_2 y, \tag{6.6}$$

where $0 < \alpha < 1, 0 < \beta < 1$. Assuming that $r_1 = r_2 = r$, $\mu_1 = \mu_2 = \mu$, and $r > \mu$, there is a steady state, $(\bar{x}, \bar{y})$, where the two species coexist:

$$r(1 - \frac{\bar{x} + \alpha \bar{y}}{K}) - \mu = 0,$$

$$r(1 - \frac{\beta \bar{x} + \bar{y}}{K}) - \mu = 0.$$

Problem 6.3. Show that the steady state of coexistence is given by

$$(\frac{K(1 - \frac{\mu}{r})(1 - \alpha)}{1 - \alpha \beta}, \frac{K(1 - \frac{\mu}{r})(1 - \beta)}{1 - \alpha \beta}).$$

and that this steady point is stable.

Problem 6.4. The model (6.5)–(6.6) with $\alpha > 1, \beta > 1$ represents the growth of two species under too aggressive competition for resources. In this case, the steady point of coexistence is given by the same expression as in Problem 6.3. Show that this steady state is unstable.

The results of Problems 6.3 and 6.4 show that when two species are using the same resources, they both will stably coexist if they do not infringe significantly upon each other, that is, if $\alpha < 1$ and $\beta < 1$, but they cannot stably coexist if the competition is very aggressive, that is, if $\alpha > 1$ and $\beta > 1$.

Cancer Model

Recall that logistic growth for a population with density x was modeled by

$$\frac{dx}{dt} = rx(1 - \frac{x}{K}) - \mu x,$$

where r is the growth rate, μ is the death rate, and K is the medium carrying capacity which is determined by the resources available to support the population. If $\mu > r$ then $\frac{dx}{dt} + (\mu - r)x \leq 0$ so that

$$x(t) \leq x(t_0)e^{-(\mu - r)t} \to 0, \quad \text{as } t \to \infty.$$

We are interested in cases where populations persist, so we shall take $\mu < r$.

If two populations x and y coexist in the same medium and follow a logistic growth, then

$$\frac{dx}{dt} = r_1 x(1 - \frac{x+y}{K}) - \mu_1 x,$$
$$\frac{dy}{dt} = r_2 y(1 - \frac{x+y}{K}) - \mu_2 y,$$

where r_1 and r_2 are the growth rates of the populations x and y, respectively, and μ_1 and μ_2 are their respective death rates. We assume that the two populations share equally the same medium, hence the term $(x+y)/K$ represents the load of the total population $x+y$ on the medium carrying capacity K. We shall apply this model to cancer in a human tissue, where x represents the density of normal healthy cells and y represents the density of cancer cells in the same tissue, and the two populations of cells are competing for space. Since cancer cells proliferate faster than normal healthy cells, we take

$$r_2 > r_1.$$

For simplicity we assume that $\mu_1 = \mu_2 = \mu$ and take $r_1 > \mu$. Writing

$$\frac{dx}{dt} = x[r_1(1 - \frac{x+y}{K}) - \mu], \tag{6.7}$$
$$\frac{dy}{dt} = y[r_2(1 - \frac{x+y}{K}) - \mu], \tag{6.8}$$

we observe that there cannot be a steady point $(\bar{x}, \bar{y})$ with $\bar{x} > 0, \bar{y} > 0$. Indeed, if such a point exists then $\frac{\bar{x}+\bar{y}}{K}$ is equal to both $1 - \frac{\mu}{r_1}$ and $1 - \frac{\mu}{r_2}$, which is impossible. The point $(0,0)$ is a steady point with Jacobian, easily computed by the factorization principle,

$$J(0,0) = \begin{pmatrix} r_1 - \mu & 0 \\ 0 & r_2 - \mu \end{pmatrix}.$$

Since both eigenvalues are positive, $(0,0)$ is an unstable node with phase portrait as displayed in Fig 3.1(A). There remains to consider the steady points

$$((1 - \frac{\mu}{r_1})K, 0) \quad \text{and} \quad (0, (1 - \frac{\mu}{r_2})K).$$

Problem 6.5. Prove that $(0, (1 - \frac{\mu}{r_2})K)$ is stable, and $((1 - \frac{\mu}{r_1})K, 0)$ is unstable.

This result means that cancer-free state is unstable whereas the steady state where all cells are cancer cells is stable.

Fig. 6.1 displays the phase portrait for the cancer model (6.7)–(6.8).

The x-nullcline is

$$\Gamma_1 : 1 - \frac{x+y}{K} = \frac{\mu}{r_1}$$

and the y-nullcline is

$$\Gamma_2 : 1 - \frac{x+y}{K} = \frac{\mu}{r_2}.$$

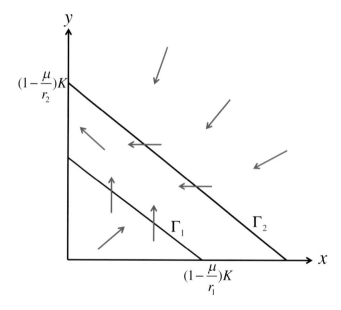

Fig. 6.1: Phase portrait for the cancer model (6.7)–(6.8).

Since $r_2 > r_1$, Γ_2 lies above Γ_1. Below Γ_1, $\frac{dx}{dt} > 0$, $\frac{dy}{dt} > 0$; above Γ_2, $\frac{dx}{dt} < 0$, $\frac{dy}{dt} < 0$; and between Γ_1 and Γ_2, $\frac{dx}{dt} < 0$ and $\frac{dy}{dt} > 0$. We see that the steady point $(0, (1 - \frac{\mu}{r_2})K)$ is globally asymptotically stable, that is, for any solution with initial values not equal to $(0,0)$ and not equal to $((1 - \frac{\mu}{r_1})K, 0)$, there holds: $(x(t), y(t)) \to (0, (1 - \frac{\mu}{r_2})K)$ as $t \to \infty$.

It is interesting to explore the dynamics of the system (6.7)–(6.8) by analysis. We have

$$\frac{d}{dt} \ln \frac{y}{x} = \frac{d}{dt}(\ln y - \ln x) = \frac{1}{y}\frac{dy}{dt} - \frac{1}{x}\frac{dx}{dt} = (r_2 - r_1)(1 - \frac{x+y}{K}). \qquad (6.9)$$

To make use of this formula we first show that if $x(0) + y(0) < K$ then for any sufficiently small $\varepsilon > 0$ with $x(0) + y(0) + \varepsilon < K$, there holds:

$$x(t) + y(t) < K - \varepsilon \quad \text{for all } t > 0. \tag{6.10}$$

Indeed, suppose this claim is not true, then there is a smallest $\bar{t}$ such that (6.10) holds for all $t < \bar{t}$, but at $t = \bar{t}$

$$x(\bar{t}) + y(\bar{t}) = K - \varepsilon. \tag{6.11}$$

It follows that

$$\frac{d}{dt}(x(t) + y(t))_{t=\bar{t}} \geq 0. \tag{6.12}$$

However, from Eqs. (6.7), (6.8) and (6.11), we get

$$\frac{d}{dt}(x(t) + y(t))|_{t=\bar{t}} \leq (K - \varepsilon)r_1(1 - \frac{K - \varepsilon}{K}) - \mu x(\bar{t})$$
$$+ (K - \varepsilon)r_2(1 - \frac{K - \varepsilon}{K}) - \mu y(\bar{t}).$$

Noting that

$$1 - \frac{K - \varepsilon}{K} = \frac{\varepsilon}{K} \quad \text{and} \quad x(\bar{t}) + y(\bar{t}) = K - \varepsilon,$$

we see that

$$\frac{d}{dt}(x(t) + y(t))|_{t=\bar{t}} \leq (K - \varepsilon)\frac{r_1 + r_2}{K}\varepsilon - \mu(K - \varepsilon)$$
$$= (K - \varepsilon)(\frac{r_1 + r_2}{K}\varepsilon - \mu) < 0$$

if $\varepsilon < \mu K/(r_1 + r_2)$, which is a contradiction to (6.12). Hence the assertion (6.10) is valid.

Substituting (6.10) into (6.9) we get

$$\frac{d}{dt} \ln \frac{y}{x} \geq (r_2 - r_1)(1 - \frac{K - \varepsilon}{K}) = \frac{(r_2 - r_1)\varepsilon}{K} \equiv \delta.$$

It follows that

$$\ln \frac{y(t)}{x(t)} \geq \ln \frac{y(0)}{x(0)} + \delta t$$

if $y(0) > 0, x(0) > 0$, so that, with $C = y(0)/x(0)$,

$$\frac{y(t)}{x(t)} \geq Ce^{\delta t}.$$

But since, by (6.10), $y(t) < K$ for all $t > 0$, we conclude that

$$x(t) \leq \frac{K}{C}e^{-\delta t} \to 0 \quad \text{as } t \to \infty. \tag{6.13}$$

From (6.8) and (6.13) we deduce that for any small $\eta > 0$ if $y(t) > (1 - \frac{\mu}{r_2})K + \eta$ and t is large, then $\frac{dy(t)}{dt} < 0$, whereas if $y(t) < (1 - \frac{\mu}{r_2})K - \eta$ and t is large then $\frac{dy(t)}{dt} > 0$. Hence $y(t) \to (1 - \frac{\mu}{r_2})K$ as $t \to \infty$.

In the above analysis we assumed that $x(0) + y(0) < K - \varepsilon$ for some small $\varepsilon > 0$. We next wish to eliminate this assumption and prove that for any initial values $(x(0), y(0))$ with $x(0) > 0$, $y(0) > 0$ it is still true that $x(t) \to 0$, $y(t) \to (1 - \frac{\mu}{r_2})K$ as $t \to \infty$. To do that all we need to show that there exist a time $t = \bar{t}$ such that

$$x(\bar{t}) + y(\bar{t}) < K - \varepsilon, \tag{6.14}$$

for then we can follow the previous arguments with initial time $t = \bar{t}$ instead of $t = 0$.

To prove (6.14) we proceed by contradiction: we assume that

$$x(t) + y(t) \geq K - \varepsilon \quad \text{for all } t > 0 \tag{6.15}$$

and derive a contradiction.

Using (6.15) in Eq. (6.7) we obtain the differential inequality

$$\frac{dx}{dt} \leq x[r_1(1 - \frac{K - \varepsilon}{K}) - \mu] = x(r_1 \frac{\varepsilon}{K} - \mu) = -\gamma x,$$

where

$$\gamma = \mu - r_1 \frac{\varepsilon}{K} > 0 \quad \text{if } \varepsilon < \frac{\mu K}{r_1}.$$

Hence

$$x(t) \leq x(0)e^{-\gamma t},$$

so that $x(t) \to 0$ as $t \to \infty$. Similarly $y(t) \to 0$ as $t \to \infty$, and this is a contradiction to (6.15).

We have thus proved:

Theorem 6.1. *The steady cancer-only state* $(0, (1 - \frac{\mu}{r_2})K)$ *is globally asymptotically stable, that is, for any initial values, $x(0) > 0$, $y(0) > 0$, $x(t) \to 0$ and $y(t) \to (1 - \frac{\mu}{r_2})K$ as $t \to \infty$.*

Thus, the model (6.7)–(6.8) predicts that, without treatment, the cancer cells will fill the entire tissue.

6.1 Numerical Simulations

To write MATLAB codes for the following problems, we can refer to Algorithms 5.1 and 5.2 in Chapter 5.

Problem 6.6. Consider the dynamical system (6.7)–(6.8) with $\mu = 1$, $r_1 = 1, 2$, $r_2 = 1, 4$, $K = 3$, and $x(0) = 2$, $y(0) = 1$, so that $x(t) > y(t)$ for t small. Simulate the system for $t > 0$ until you arrive at time T such that $y(T) = x(T)$.

Problem 6.7. There are anti-cancer drugs that decrease the death rate of cancer cells but not of normal healthy cells. Under a treatment with such a drug, Eq. (6.8) for cancer cells becomes

$$\frac{dy}{dt} = y[r_2(1 - \frac{x+y}{K}) - \sigma\mu], \quad \sigma > 1, \tag{6.16}$$

where σ represents the effect of the drug. Repeat the calculations of Problem 6.6 for the system (6.7), (6.16) and $\mu = 1$, $r_1 = 1.2$, $r_2 = 1.4$, $k(0) = 3$ and $x(0) = 5$, $y(0) = 1$ until you arrive at time $T = T(\sigma)$ such that $y(T) = x(T)$; do it for $\sigma = 1, 1.1, 1.2, 1.3, 1.4, 1.5, 1.6, 1.7, 1.8, 1.9$, and 2.0, and draw the approximate curve $T = T(\sigma)$ for $1 < \sigma < 2$.

6.1.1 Revisiting Euler Method for Solving ODE – Backward Euler Method

In the numerical section of Chapter 5, we have introduced forward Euler method. It is an explicit method, which is easy to implement, for both scalar and system problems. We have also seen that it is first order accurate, for which we may have to use a small time step to reach a reasonable accuracy. Moreover, the method carries a time step constraint, regardless of accuracy, in order to be stable. If the time step exceeds the constrained value, the solution will blow up. For some practical problems, this time step constraint is prohibitive in terms of computational time. Therefore, it is desirable to get rid of this constraint for these cases.

Recall the differential equation in (5.20),

$$\frac{dx}{dt} = f(x,t), \quad t \geq t_0, \quad x(t_0) = x_0,$$

and Eq. (2.23),

$$x(t) = x(t_n) + \int_{t_n}^{t} f(x, \tau)d\tau, \quad t_n \geq t_0,$$

where t_n is some time point. If we approximate the integral by $hf(x(t_n + h), t_n + h)$ then we end up with the **backward Euler method**, a different method from the forward Euler method introduced in Chapter 5. This leads to the following formula

$$X_{n+1} = X_n + hf(X_{n+1}, t_{n+1}), \tag{6.17}$$

where the notation is the same as in Section 5.1. Note that the difference between forward Euler and backward Euler is that we are using unknown X_{n+1} in function f of Eq. (6.17). To solve Eq. (6.17), one needs to solve

$$X_{n+1} - hf(X_{n+1}, t_{n+1}) = X_n,$$

which may require a nonlinear solver if f is nonlinear. Recall that in the forward Euler method, X_{n+1} is directly computed from the right-hand side using X_n, which makes it very simple to implement and calculate.

Why would we want to use an implicit method which involves possibly time consuming nonlinear solvers? Let's consider again the test equation $y' = \lambda y$, $\lambda < 0$. We apply the backward Euler method to the test problem, and get the formula

$$X_{n+1} = X_n + h\lambda X_{n+1},$$

and therefore

$$(1 - h\lambda)X_{n+1} = X_n.$$

The linearity of this problem leads to a simple solution of the implicit method, namely,

$$X_{n+1} = \frac{X_n}{1 - h\lambda}.$$

Because we assumed that $\lambda < 0$, we have $|X_{n+1}| < |X_n|$ regardless of the choice of h. In other words, the numerical solution decreases in magnitude and will never blow up, and hence it is stable for any time step h. We call this numerical approximation 'unconditionally stable.'

In general, problems that require very small time step h with explicit methods due to rapid variation in the solution, are called **stiff problems**. When we encounter stiff problems, implicit methods become very useful since they avoid very small time steps, which means time consuming simulations. In that case, solving a nonlinear problem at each time step will pay off by gaining stability.

Problem 6.8. Consider the test problem with $\lambda = 20$ and $y_0 = 1$. The sample MATLAB codes are in Algorithm 6.1. Test $h = 0.01, 0.05, 0.1, 0.2, 0.5$. What do you observe? Compare the result with that of Problem 5.7.

Algorithm 6.1. Backward Euler method for the test problem (backward_Euler.m)

```
% This is a code to solve dy/dt = lambda*y, t in [0,1], y(0)=1
% (lambda=-20) using backward Euler method
lambda = -20;
h = .5;
t = 0:h:1;
Nt = length(t);
y = zeros(1,Nt); % preset y as a zero matrix with the same length
   as t
y(1)=1; % initial condition

for i = 1:(Nt-1)
   y(i+1) = y(i)/(1-lambda*h);
end

plot(t,y,'-o')
```

Problem 6.9. Implement the backward Euler method for

$$\frac{dy}{dt} = -y + t, \quad y(0) = 1, \quad 0 \le t \le 1.$$

(a) Derive the exact solution. (b) Compare your numerical solution ($h = 0.01, 0.005$, 0.001) with the exact solution by plotting them together.

Problem 6.10. Implement the backward Euler method for

$$\frac{dy}{dt} = -y^2 + t, \quad y(0) = 1, \quad 0 \le t \le 1.$$

Use $h = 0.01, 0.005, 0.001$. [Hint: you have to solve a quadratic equation at each time step.]

In MATLAB, explicit methods 'ode45' or others such as 'ode23' are often used when solving nonstiff problems. There are also implicit methods, such as 'ode15s,' that would efficiently and robustly calculate stiff problems. Other choices can be found by searching the MATLAB library.

Chapter 7
General Systems of Differential Equations

In this chapter, we develop a theory for a system of differential equations that will be used to study models with many species. We write the system either in the form

$$\frac{dx_i}{dt} = f_i(x_1, x_2, \cdots, x_n), \quad i = 1, 2, \cdots, n \tag{7.1}$$

or, in vector notation,

$$\frac{d\mathbf{x}}{dt} = \mathbf{f}(\mathbf{x}), \tag{7.2}$$

where $\mathbf{x} = (x_1, \cdots, x_n), \mathbf{f} = (f_1, \cdots, f_n)$.

In order to understand the behavior of solutions of system (7.1) with f_i that are general nonlinear functions of $x_1, \cdots, x_n$, we begin with a study of linear systems

$$\frac{dx_i}{dt} = \sum_{j=1}^{n} a_{ij} x_j, \quad 1 \leq i \leq n. \tag{7.3}$$

In the sequel we shall need some results from Linear Algebra, which extend Theorems 3.1 and 3.2 to systems

$$\sum_{j=1}^{n} \alpha_{ij} x_j = b_j, \quad 1 \leq i \leq n \tag{7.4}$$

for unknowns $x_1, \cdots, x_n$, where α_{ij} and b_j are given numbers, and $n \geq 2$.

Electronic supplementary material The online version of this chapter (doi: 10.1007/978-3-319-29638-8_7) contains supplementary material, which is available to authorized users.

C.-S. Chou, A. Friedman, *Introduction to Mathematical Biology*,
Springer Undergraduate Texts in Mathematics and Technology,
DOI 10.1007/978-3-319-29638-8_7

We introduce the matrix

$$A = \begin{pmatrix} \alpha_{11} & \alpha_{12} & \cdots & \alpha_{1n} \\ \alpha_{21} & \alpha_{22} & \cdots & \alpha_{2n} \\ \cdots & & & \\ \cdots & & & \\ \alpha_{n1} & \alpha_{n2} & \cdots & \alpha_{nn} \end{pmatrix}, \text{ or } A = (\alpha_{ij}),$$

and its determinant

$$\det A = \begin{vmatrix} \alpha_{11} & \alpha_{12} & \cdots & \alpha_{1n} \\ \alpha_{21} & \alpha_{22} & \cdots & \alpha_{2n} \\ \cdots & & & \\ \cdots & & & \\ \alpha_{n1} & \alpha_{n2} & \cdots & \alpha_{nn} \end{vmatrix}.$$

Theorem 7.1. *(i) If $\det A \neq 0$ then the system (7.4) has a unique solution $(x_1, \cdots, x_n)$ for any prescribed vector $(b_1, \cdots, b_n)$, and the solution is given by*

$$x_i = \frac{1}{\det A} \begin{vmatrix} \alpha_{11} & \cdots & \alpha_{1,i-1} & b_1 & \alpha_{1,i+1} & \cdots & \alpha_{1n} \\ \alpha_{21} & \cdots & \alpha_{2,i-1} & b_2 & \alpha_{2,i+1} & \cdots & \alpha_{2n} \\ \cdots & & & & & & \\ \cdots & & & & & & \\ \alpha_{n1} & \cdots & \alpha_{n,i-1} & b_n & \alpha_{n,i+1} & \cdots & \alpha_{nn} \end{vmatrix};$$

in particular, the only solution of the homogeneous system

$$\sum_{j=1}^{n} \alpha_{ij} x_j = 0, \quad 1 \leq i \leq n \tag{7.5}$$

is the zero solution $(x_1, \cdots, x_n) = (0, \cdots, 0)$;
(ii) If $\det A = 0$ then there exist nonzero solutions $(x_1, \cdots, x_n)$ of the homogeneous system (7.5).

Vectors $\mathbf{w}_1, \cdots, \mathbf{w}_n$ are said to be **linearly dependent** if there exists a nonzero vector $\mathbf{c} = (c_1, \cdots, c_n)$ such that

$$\sum_{j=1}^{n} c_j \mathbf{w}_j = 0. \tag{7.6}$$

Conversely, vectors $\mathbf{w}_1, \cdots, \mathbf{w}_n$ are **linearly independent** if a relation of the form (7.6) can only be satisfied when $c_1 = \cdots = c_n = 0$.

It will be useful to express Theorem 7.1 in vector notation, with vectors

$$\boldsymbol{\alpha}_j = \begin{pmatrix} \alpha_{1j} \\ \alpha_{2j} \\ . \\ . \\ . \\ \alpha_{nj} \end{pmatrix}, \quad \mathbf{b} = \begin{pmatrix} b_1 \\ b_2 \\ . \\ . \\ . \\ b_n \end{pmatrix}.$$

Theorem 7.2. *(i) If* $\det A \neq 0$ *then any vector* **b** *can be represented in a unique way as a linear combination of the vectors* $\boldsymbol{\alpha}_j$, *that is,*

$$\mathbf{b} = \sum_{j=1}^{n} x_j \boldsymbol{\alpha}_j,$$

where $x_1, \cdots, x_n$ *are uniquely determined by* **b** *(by Theorem 7.1(ii)); in particular, if* $\mathbf{b} = \mathbf{0}$ *then* $x_1 = \cdots = x_n = 0$, *hence the vectors* $\boldsymbol{\alpha}_1, \cdots, \boldsymbol{\alpha}_n$ *are linearly independent.*

(ii) If $\det A = 0$ *then there are nonzero vectors* $(x_1, \cdots, x_n)$ *such that*

$$\sum_{j=1}^{n} x_j \boldsymbol{\alpha}_j = 0,$$

that is, the vectors $\boldsymbol{\alpha}_1, \cdots, \boldsymbol{\alpha}_n$ *are linearly dependent.*

The proofs of Theorems 7.1 and 7.2 can be found, for instance, in Reference [3].

Corollary 7.1. *If n-vectors* $\mathbf{w}_1, \cdots, \mathbf{w}_n$ *are linearly independent, then any n-vector* $\mathbf{x}_0$ *can be expressed as a linear combination*

$$\mathbf{x}_0 = \sum_{j=1}^{n} c_j \mathbf{w}_j$$

of $\mathbf{w}_1, \cdots, \mathbf{w}_n$ *and the coefficients* $c_1, \cdots, c_n$ *are uniquely determined.*

Following the case $n = 2$ we seek a solution of the system (7.3) in the form

$$x_i = v_i e^{\lambda t}, \quad 1 \leq i \leq n.$$

Then λ and the v_i satisfy the equations $\lambda v_i = \sum_{j=1}^{n} a_{ij} v_j$, or

$$\sum_{j=1}^{n} (a_{ij} - \lambda \delta_{ij}) v_j = 0, \quad 1 \leq i \leq n, \tag{7.7}$$

where $\delta_{ij} = 0$ if $i \neq j$, $\delta_{ij} = 1$ if $i = j$. By Theorem 7.1 this system has a nonzero solution if and only if

$$\det(a_{ij} - \lambda \delta_{ij}) = 0, \quad \text{or} \quad \det(A - \lambda I) = 0, \tag{7.8}$$

where I is the unit matrix (with elements δ_{ij}). Equation (7.8) is a polynomial equation of order n,

$$\lambda^n + a_1 \lambda^{n-1} + \cdots + a_{n-1} \lambda + a_n = 0; \tag{7.9}$$

it is called the **characteristic equation** of A and its zeros are called **eigenvalues** of A. If λ is an eigenvalue then a nonzero solution of (7.7) is called an **eigenvector** corresponding to eigenvalue λ.

It is known that a polynomial equation of order n has n zeros. If all the eigenvalues $\lambda_1, \cdots, \lambda_n$ of Eq. (7.9) are distinct, then we can associate them with n different solutions of the system (7.3), namely,

$$e^{\lambda_1 t}\mathbf{w}_1, e^{\lambda_2 t}\mathbf{w}_2, \cdots, e^{\lambda_n t}\mathbf{w}_n, \tag{7.10}$$

where $\mathbf{w}_i$ is an eigenvector associated with λ_i.

Theorem 7.3. *If all the λ_j are distinct then $\mathbf{w}_1, \mathbf{w}_2, .., \mathbf{w}_n$ are linearly independent, that is, if $c_1, .., c_n$ are any numbers such that*

$$c_1\mathbf{w}_1 + c_2\mathbf{w}_2 + \cdots + c_n\mathbf{w}_n = 0, \tag{7.11}$$

then $c_1 = c_2 = \cdots = c_n = 0$.

Proof. Note that $A^2\mathbf{w}_j = A(A\mathbf{w}_j) = A(\lambda_j\mathbf{w}_j) = \lambda_j A\mathbf{w}_j = \lambda_j^2\mathbf{w}_j$; similarly $A^3\mathbf{w}_j = A(A^2\mathbf{w}_j) = \lambda_j^2 A\mathbf{w}_j = \lambda_j^3\mathbf{w}_j$, and, by induction, $A^k\mathbf{w}_j = \lambda_j^k\mathbf{w}_j$ for any positive integer k. Applying the matrix A^k to Eq. (7.11) we get,

$$\sum_{j=1}^n \lambda_j^k c_j\mathbf{w}_j = 0, \quad 0 \le k \le n-1.$$

or, if $\mathbf{w}_j = (v_{j1}, \cdots, v_{jn})$,

$$\sum_{j=1}^n \lambda_j^k (c_j v_{jm}) = 0, \quad 0 \le k \le n-1$$

for any m, $1 \le m \le n$. But the matrix (λ_j^k) is the Vandermonde matrix whose determinant is

$$\prod_{0 \le i \le j \le n-1} (\lambda_j - \lambda_i),$$

and it is thus not equal to zero since all the λ_j are distinct. Hence, by Theorem 7.1, $c_j v_{jm} = 0$ for all j. Since this is true for any m, $c_j\mathbf{w}_j = 0$ and hence $c_j = 0$.

From Theorem 7.1 and Corollary 7.1 we conclude that any vector $\mathbf{x}_0 = (x_{01}, \cdots, x_{0n})$ can be written as linear combination

$$\mathbf{x}_0 = c_1\mathbf{w}_1 + c_2\mathbf{w}_2 + \cdots + c_n\mathbf{w}_n$$

with uniquely determined coefficients $c_1, \cdots, c_n$. Hence there exists a unique solution of (7.3) with initial conditions

$$x_i(0) = x_{i0}, \quad 1 \le i \le n \tag{7.12}$$

in the form

$$\mathbf{x}(t) = \sum c_i\mathbf{w}_i e^{\lambda_i t}. \tag{7.13}$$

So far we assumed that the λ_i are all distinct. Suppose next that some of the eigenvalues are equal; for example, $\lambda_1 = \lambda_2 = \lambda_3$. Then in addition to $\mathbf{w}_1 e^{\lambda_1 t}$ we can also find two other solutions $\mathbf{w}_2(t) e^{\lambda_1 t}$, $\mathbf{w}_3(t) e^{\lambda_1 t}$ where $\mathbf{w}_2(t), \mathbf{w}_3(t)$ have the form

$$\mathbf{w}_2(t) = t\mathbf{w}_1 + \mathbf{w}_{22}, \quad \mathbf{w}_3(t) = t^2 \mathbf{w}_1 + t\mathbf{w}_{32} + \mathbf{w}_{33} \tag{7.14}$$

so that $\mathbf{w}_1, \mathbf{w}_2(0), \mathbf{w}_3(0), \mathbf{w}_4, \cdots, \mathbf{w}_n$ are linearly independent, and hence the general solution of (7.3), (7.12) is still given by (7.13), but with $\mathbf{w}_2, \mathbf{w}_3$ of the form (7.14).

The above considerations extend to the general case where some eigenvalues are equal to each other. We can then conclude that:

If $Re(\lambda_j) < 0$ for $1 \le j \le n$, then any solution of (7.3) satisfies: $x(t) \to 0$ as $t \to \infty$. (7.15)

We can now proceed with the general system (7.1). As in Theorem 2.1, if the derivatives $\partial f_i / \partial x_j$ are continuous functions for all $\mathbf{x}$, then for any initial values $(x_{10}, \cdots, x_{n0})$ there exists a unique solution $(x_1(t), \cdots, x_n(t))$ of (7.1) satisfying the initial conditions

$$x_j(0) = x_{j0}, \quad 1 \le j \le n,$$

for all $t > 0$ as long as the solution remains bounded. We refer to the solution as a **trajectory**. The solution can also be extended to $t < 0$.

A point $\bar{\mathbf{x}} = (\bar{x}_1, \cdots, \bar{x}_n)$ such that $\mathbf{f}(\bar{\mathbf{x}}) = \mathbf{0}$ is called an **equilibrium point**, a **stationary point**, or a **steady point** of the system (7.1). If $\bar{\mathbf{x}}$ is a steady point, then the unique trajectory $\mathbf{x}(t)$ with $\mathbf{x}(0) = \bar{\mathbf{x}}$ is $\mathbf{x}(t) \equiv \bar{\mathbf{x}}$, for all $t \ge 0$.

Writing

$$\mathbf{f}_i(\mathbf{x}) = \mathbf{f}_i(\bar{\mathbf{x}}) + \sum_{j=1}^n (x_j - \bar{x}_j) \left[\frac{\partial f_i}{\partial x_j} + \varepsilon_j(|\mathbf{x} - \bar{\mathbf{x}}|) \right],$$

where $\varepsilon_j(s) \to 0$ if $s \to 0$, we see that the linear system of differential equations

$$\frac{dx_i}{dt} = \sum_{j=1}^n a_{ij}(x_j - \bar{x}_j), \quad (a_{ij} = \frac{\partial f_i(\bar{\mathbf{x}})}{\partial x_j}) \tag{7.16}$$

is a good approximation to (7.1) near $\mathbf{x} = \bar{\mathbf{x}}$. The matrix

$$J = \begin{pmatrix} \frac{\partial f_1}{\partial x_1} & \frac{\partial f_1}{\partial x_2} & \cdots & \frac{\partial f_1}{\partial x_n} \\ & \vdots & & \\ \frac{\partial f_n}{\partial x_1} & \frac{\partial f_n}{\partial x_2} & \cdots & \frac{\partial f_n}{\partial x_n} \end{pmatrix},$$

where $\frac{\partial f_i}{\partial x_j}$ is computed at $\bar{\mathbf{x}}$, is called the **Jacobian matrix** at $\bar{\mathbf{x}}$; we also write $J = (\frac{\partial f_i}{\partial x_j})$. As in the analysis in Chapters 2 and 3, we wish to determine under what conditions all solutions of (7.1), or (7.2), with initial values near $\bar{\mathbf{x}}$ converge to $\bar{\mathbf{x}}$ as $t \to \infty$, and in this case we call $\bar{\mathbf{x}}$ a **stable** equilibrium point, or, more precisely, **asymptotically stable** equilibrium point.

Since the linear system (7.16) is a good approximation to the full system (7.1) near the point $\bar{\mathbf{x}}$, using the representation (7.13) of the general solution of (7.16), and recalling (7.15), we have the following result:

Theorem 7.4. *If $Re(\lambda_j) < 0$ for each eigenvalue of the Jacobian matrix at $\bar{\mathbf{x}}$, then the point $\bar{\mathbf{x}}$ is an asymptotically stable (or, briefly, a stable) equilibrium point for (7.1).*

This means that any trajectory $\mathbf{x}(t)$, with $\mathbf{x}(0)$ near $\bar{\mathbf{x}}$, converges to $\bar{\mathbf{x}}$ as $t \to \infty$.

The next question is under what conditions on the coefficients $a_1, a_2, \cdots, a_n$ of the characteristic polynomial (7.9) is it true that $Re(\lambda_j) < 0$ for all j. The answer is provided by the well-known criteria of **Routh-Hurwitz** which actually holds for any polynomial (7.9). In the sequel we shall need to use the Routh-Hurwitz criterion only in case $n = 3$:

Theorem 7.5. *All the roots of a polynomial*

$$\lambda^3 + a_1 \lambda^2 + a_2 \lambda + a_3 = 0 \qquad (7.17)$$

have negative real parts if and only if $a_1 > 0, a_3 > 0$, and $a_1 a_2 > a_3$.

The proof of Theorem 7.5 and the general Routh-Hurwitz theorem can be found in Reference [4]. Theorem 7.5 will be used in the following example.

Example 7.1. Consider the model of one predator x and two prey species y and z:

$$\frac{dx}{dt} = \beta_1 xy + \beta_2 xz - \mu x,$$

$$\frac{dy}{dt} = r_1 y - \gamma_1 xy,$$

$$\frac{dz}{dt} = r_2 z(1 - z) - \gamma_2 xz.$$

It is easily seen that the only steady point $(\bar{x}, \bar{y}, \bar{z})$ with $\bar{x} > 0, \bar{y} > 0, \bar{z} > 0$ is given by

$$\bar{x} = \frac{r_1}{\gamma_1}, \quad \bar{z} = 1 - \frac{\gamma_2}{r_2}\bar{x}, \quad \beta_1 \bar{y} = \mu - \beta_2 \bar{z}$$

provided $\gamma_2 \bar{x} < r_2$ and $\beta_2 \bar{z} < \mu$. To check whether this point is stable we compute the Jacobian matrix $J(\bar{x}, \bar{y}, \bar{z})$ using the factorization rule. We find that

$$J(\bar{x}, \bar{y}, \bar{z}) = \begin{pmatrix} 0 & \beta_1 \bar{x} & \beta_2 \bar{x} \\ -\gamma_1 \bar{y} & 0 & 0 \\ -\gamma_2 \bar{z} & 0 & -r_2 \bar{z} \end{pmatrix}.$$

We then compute that the characteristic equation has the form (7.17) with

$$a_1 = r_2 \bar{z}, \quad a_2 = \beta_1 \gamma_1 \bar{x}\bar{y} + \beta_2 \gamma_2 \bar{x}\bar{z}, \quad a_3 = \beta_1 \gamma_1 r_2 \bar{x}\bar{y}\bar{z} \quad \text{and} \quad a_1 a_2 = a_3 + \beta_2 \gamma_2 r_2 \bar{x}\bar{z}^2 > a_3.$$

Hence, by the Routh-Hurwitz criterion the steady point $(\bar{x}, \bar{y}, \bar{z})$ is stable.

Problem 7.1. Consider the model of one predator x and two prey species y and z:

$$\frac{dx}{dt} = r_1 x(1 - \frac{x}{A}) + \beta_1 xy + \beta_2 xz - \mu x,$$

$$\frac{dy}{dt} = r_1 y - \gamma_1 xy,$$

$$\frac{dz}{dt} = r_2 z(1 - z) - \gamma_2 xz.$$

Compute the unique steady point $(\bar{x}, \bar{y}, \bar{z})$ with $\bar{x} > 0, \bar{y} > 0, \bar{z} > 0$ and use the Routh-Hurwitz theorem to prove that $(\bar{x}, \bar{y}, \bar{z})$ is stable.

Consider a model of two predators, x and y, and one prey, z:

$$\frac{dx}{dt} = r_1 x(1 - \frac{x}{k_1}) + \beta_1 xz,$$

$$\frac{dy}{dt} = r_2 y(1 - \frac{y}{k_2}) + \beta_2 yz, \tag{7.18}$$

$$\frac{dz}{dt} = \alpha z(1 - \frac{z}{B}) - r_1 xz - r_2 yz.$$

Note that in this model each of the predators, x and y, can actually survive on its own, even if they do not feed on z.

Problem 7.2. Show that the system (7.18) has a unique steady point $(\bar{x}, \bar{y}, \bar{z})$ with $\bar{x} > 0, \bar{y} > 0, \bar{z} > 0$, and that this point is stable.

Problem 7.3. Consider a model of one prey (x) and two predators (y_i):

$$\frac{dx}{dt} = ax(1 - \frac{x}{A}) - \sum_{j=1}^{2} bxy_j,$$

$$\frac{dy_i}{dt} = -c_i y_i + d_i xy_i, \quad i = 1, 2,$$

where $\frac{c_1}{d_1} < \frac{c_2}{d_2} < A$. There are four equilibrium points:

$$(0,0,0), \quad (A,0,0), \quad (\frac{c_1}{d_1}, \frac{a}{b}(1 - \frac{c_1}{Ad_1}),0), \quad (\frac{c_2}{d_2},0,\frac{a}{b}(1 - \frac{c_2}{Ad_2})).$$

Determine which of these points are stable.

7.1 Numerical Simulations

7.1.1 Solving for the Steady States

In the previous chapters we discussed how to solve an ordinary differential equation

$$\frac{d\mathbf{x}}{dt} = \mathbf{f}(\mathbf{x})$$

by using Euler's method or the function ode45 in MATLAB. For many problems, we need to find the steady states of the system and calculate the local stability, as seen in this and previous chapters. The steady states are not always solvable analytically, and sometimes we need to rely on computational tools to approximate those steady states. Here, we introduce a method of how to solve for the stationary solution of an ODE system, i.e., the solution of the steady state equation

$$\mathbf{f}(\mathbf{x}) = \mathbf{0}.$$

If $\mathbf{f}(\mathbf{x})$ is linear or is of a simple function form, this equation may be solved analytically; however, if it is nonlinear or if the system is large, solving by hand is not feasible, and thus one needs to use **root-finding algorithms**. Let us start with f being a real-valued function, and

$$f(x) = 0. \tag{7.19}$$

Two of the best known root finding algorithms for (7.19) are the *bisection method* and *Newton's method*, the latter named after the eminent 17th century mathematician and scientist Isaac Newton. The bisection method is a 'gradient free' approach and usually takes longer to converge but it is more robust (you can always find a root if the initial interval is valid, that is, if it contains at least one root). Newton's method uses gradient (slope in one dimension) information and is more efficient; however, it may fail for certain problems or initial guesses.

7.1.2 Bisection Method

The idea of the bisection method comes from the intermediate value theorem: continuous function f must have at least one root in the interval (a,b) if $f(a)$ and $f(b)$ have opposite signs. The method repeatedly bisects an interval then selects a subinterval in which a root must lie (the function values at the two ends of the subinterval have opposite signs). Suppose that we have two initial points $a_0 = a$ and $b_0 = b$ such that $f(a)f(b) < 0$. The method divides the interval into two by computing the midpoint $c = \frac{a+b}{2}$ of the interval. If c is a root, that is $f(c) = 0$, then the algorithm terminates. Otherwise, the algorithm checks $f(a)f(c)$ and $f(c)f(b)$, one of which must be negative. If $f(a)f(c) < 0$, the root must lie in the interval (a,c) and the method sets a as a_1 and c as b_1. If $f(c)f(b) < 0$, the root must lie in the interval (c,b) and the method sets c as a_1 and b as b_1. Repeating this process, we can construct a sequence of intervals $[a_n, b_n]$ such that

$$|b_n - a_n| = \frac{|b_0 - a_0|}{2^n}.$$

Since the root must lie in these subintervals, the best estimate for the location of the root is the midpoint of the smallest subinterval found. In that case, the absolute error after n steps is at most

$$\frac{|b - a|}{2^{n+1}}. \tag{7.20}$$

If either endpoint of the interval is used for estimate of the root, then the maximum absolute error is

$$\frac{|b - a|}{2^n}. \tag{7.21}$$

With this method, it is also easy to estimate the number of iterations in order to reach a certain level of error. For example, if we use the midpoint to estimate the root and want the error to be around ε. Then the step N such that the error reaches ε is

$$\frac{|b - a|}{2^{N+1}} \approx \varepsilon,$$

and therefore

$$N \approx \log_2 \frac{|b - a|}{\varepsilon}.$$

A sample code with tolerance 10^{-5} to solve $x^3 = 1.5$ in the interval $[0, 2]$ is shown in Algorithm 7.1.

Problem 7.4. Modify the sample code in Algorithm 7.1 to solve $x^4 = 15$ in the interval $[0, 2]$, with tolerance 10^{-5}. Set a counter in the while loop in the code to determine how many iterations needed to reach the tolerance 10^{-5}. Plot the convergence history, that is, iterations versus $f(x)$, with x being the approximate root.

Algorithm 7.1. Bisection method (bisection_method.m)

```
%%% This code is using bisection method to solve the root of
%%% x^3 = 1.5 in the interval [0,2].

a = 0;
b = 2;
TOL = 10^-5;   % tolerance to stop the iterations
f = @(x)(x^3-1.5);

while ( b-a >= TOL )
    m = (a+b)/2;   % midpoint of the intervals
    if ( abs(f(m))<1e-10 )  % when the function value is
        break;             %   close enough to zero
    elseif ( f(a)*f(m)<0 )
        b = m;
    elseif ( f(m)*f(b)<0 )
        a = m;
    end
    m   % print out the approximation (using midpoint of
        % subintervals)
end
```

7.1.3 Newton's Method

While the bisection method only uses the sign of the function value $f(x)$ in Eq. (7.19), Newton's method uses more information of the function, including the function values and derivatives. Given an initial guess x_0, Newton's method generates a sequence of approximations of the root, $\{x_n\}$, by

$$x_{n+1} = x_n - \frac{f(x_n)}{f'(x_n)}, \quad n = 0, 1, \ldots, \tag{7.22}$$

where x_0 stands for an initial guess. This idea originates from the linear approximation near the root: if we start with an initial guess close enough to the true root, we can use the linear approximation for the tangent line of the function, and the x-intercept will typically be a better approximation for the true root. Thus, starting at a certain guess x_n, with function value $f(x_n)$, the tangent line passing that point is

$$y = f'(x_n)(x - x_n) + f(x_n),$$

and the x-intercept of this line, denoted by x_{n+1}, satisfies

$$0 = f'(x_n)(x_{n+1} - x_n) + f(x_n).$$

Hence, the next approximation of the root, x_{n+1}, is

$$x_{n+1} = x_n - \frac{f(x_n)}{f'(x_n)},$$

assuming that $f'(x_{x_n})$ is nonzero. If we do this iteratively, the sequence usually converges pretty quickly to the true root. In Algorithm 7.2, we show how to implement Newton's method to solve $x^2 = 23$.

Newton's method can be easily extended to solve the general nonlinear system

$$\mathbf{f}(\mathbf{x}) = \mathbf{0}, \quad \mathbf{x} \in R^n,$$

where $\mathbf{f}$ is a vector-valued function from R^n to R^n.

The formula (7.22) becomes

$$\mathbf{x}_{n+1} = \mathbf{x}_n - \left[\mathbf{f}'(\mathbf{x}_n)\right]^{-1} \mathbf{f}(\mathbf{x}_n). \tag{7.23}$$

where $\mathbf{f}'(\mathbf{x}_n)$ is the $n \times n$ Jacobian matrix of $\mathbf{f}$ evaluated at $\mathbf{x}_n$.

Comparing (7.22) with (7.23), note that we simply replaced the division of $f'(x_n)$ in (7.22) by left multiplication by the inverse of the $n \times n$ Jacobian matrix. In practice, it is more common to rewrite (7.23) in the form

$$\mathbf{f}'(\mathbf{x}_n)(\mathbf{x}_{n+1} - \mathbf{x}_n) = \mathbf{f}(\mathbf{x}_n),$$

which is a linear system. Define $\mathbf{e} = \mathbf{x}_{n+1} - \mathbf{x}_n$, we can use standard linear solvers to solve $\mathbf{e}$ in

$$\mathbf{f}'(\mathbf{x}_n)\mathbf{e} = \mathbf{f}(\mathbf{x}_n)$$

and obtain the approximation of the next step by

$$\mathbf{x}_{n+1} = \mathbf{e} + \mathbf{x}_n.$$

Algorithm 7.2. Newton's method to solve a scalar equation (newtons_method.m)

```
% This is the code of Newton's method for solving
% f(x)=x^2-23=0, with x=x0 as initial guess

TOL = 1e-10;   % tolerance of the function value
               % (ideally a small number)
x0 = 2;
x  = x0;
y  = x^2 - 23;

yvec = [];  %  preset an empty matrix for later use
iter = 0;   %  counter of iterations
while abs(y) > TOL,
    y = x^2 - 23;
    x = x - y/(2*x^1); % f'(x_n) is derived by hand a priori
    iter = iter + 1;
    yvec = [yvec, y];  % record the convergence history
end

display(x)
iter
plot(yvec,'-o')
```

Problem 7.5. Modify Algorithm 7.2 and implement Newton's method to solve $x^5 = 213$. Use an initial guess $x_0 = 2$. What is the root you find? How many iterations do you need to reach the tolerance 10^{-12}. Plot the convergence history.

Chapter 8
The Chemostat Model Revisited

In Chapter 2 we considered the chemostat model and used mathematics to answer the question: How should we choose the outflow rate in order to harvest the maximum amount of bacteria. Our model however was incomplete because we assumed that the nutrient concentration in the growth chamber is constant in time, and hence our answer is questionable. In the present chapter we want to correct the answer, by basing it on a more complete mathematical model of the chemostat.

We begin by introducing the following notation:

$$V = \text{volume of the bacterial chamber,}$$
$$C(t) = \text{concentration of nutrients in the chamber,}$$
$$C_0 = \text{constant concentration of nutrients supply,}$$
$$r = \text{rate of inflow and outflow,}$$
$$B(t) = \text{concentration of the bacteria in the chamber.}$$

We assume that

$$\frac{\text{mass of the bacteria formed}}{\text{mass of the nutrients used}} = constant = \gamma;$$

γ is the *yield constant*; $\gamma < 1$ since bacteria of mass 1 is formed by consumption of nutrients of larger mass $1/\gamma$. By conservation of nutrient mass,

$$\text{rate of change} = \text{input} - \text{washout} - \text{consumption.}$$

Based on experimental evidence we take the rate of bacterial growth in the entire bacterial chamber to be

$$\frac{m_0 C}{a+C} B,$$

Electronic supplementary material The online version of this chapter (doi: 10.1007/978-3-319-29638-8_8) contains supplementary material, which is available to authorized users.

© Springer International Publishing Switzerland 2016
C.-S. Chou, A. Friedman, *Introduction to Mathematical Biology*,
Springer Undergraduate Texts in Mathematics and Technology,
DOI 10.1007/978-3-319-29638-8_8

where m_0 and a are constants, and then the rate of nutrient consumption in the entire chamber is

$$\frac{m_0 C}{a+C}\frac{B}{\gamma},$$

since mass $1/\gamma$ of the bacteria is formed from consumption of mass 1 of nutrients. By conservation of nutrient mass,

$$\frac{d}{dt}(VC) = C_0 r - Cr - \frac{m_0 C}{a+C}\frac{B}{\gamma},$$

where $C = C(t), B = B(t)$. Dividing both sides by V and setting $D = r/V$ (the **dilution rate**), we get

$$\frac{dC}{dt} = (C_0 - C)D - \frac{mC}{a+C}\frac{B}{\gamma}, \tag{8.1}$$

where $m = m_0/V$. Similarly, by conservation of bacterial mass in the entire chamber,

$$\frac{d}{dt}(VB) = \frac{m_0 C}{a+C}B - Br.$$

Dividing both sides by V, we get

$$\frac{dB}{dt} = B\left(\frac{mC}{a+C} - D\right). \tag{8.2}$$

The units of C_0, C, a, and B are mass/volume (e.g., g/cm^3), and the units of m and D are 1/time (e.g., 1/sec); γ is a dimensionless parameter. The parameter m_0 is a consumption rate in the entire chamber with volume V, and m is the consumption rate in the mass/volume units of C and B.

We can simplify the system (8.1)–(8.2) by scaling, taking

$$c = \frac{C}{C_0}, \quad b = \frac{B}{\gamma C_0}, \quad \bar{t} = Dt.$$

Introducing new parameters $\bar{m} = \frac{m}{D}, \bar{a} = \frac{a}{C_0}$, the system (8.1) and (8.2) takes the following simplified form:

$$\frac{dc}{d\bar{t}} = 1 - c - \frac{\bar{m}c}{\bar{a}+c}b, \tag{8.3}$$

$$\frac{db}{d\bar{t}} = b\left(\frac{\bar{m}c}{\bar{a}+c} - 1\right). \tag{8.4}$$

We see that $(c,b) = (1,0)$ is a steady state. The Jacobian matrix at $(1,0)$ is

$$J(1,0) = \begin{pmatrix} -1 & -\frac{\bar{m}}{\bar{a}+1} \\ 0 & \frac{\bar{m}}{\bar{a}+1} - 1 \end{pmatrix}$$

and the eigenvalues are $\lambda_1 = -1, \lambda_2 = \frac{\bar{m}}{\bar{a}+1} - 1$. Hence $(1,0)$ is stable if $\lambda_2 < 0$ and unstable if $\lambda_2 > 0$. Another steady state (c_*, b_*) is obtained by solving (from Eq. (8.4))

$$\frac{\bar{m}c}{\bar{a}+c} = 1, \quad \text{i.e., } c = \frac{\bar{a}}{\bar{m}-1}$$

and then (from Eq. (8.3))

$$1 - c - b = 0, \quad \text{i.e., } b = 1 - \frac{\bar{a}}{\bar{m}-1}.$$

This steady point is biologically relevant only if

$$\bar{m} > 1 \quad \text{and} \quad \frac{\bar{a}}{\bar{m}-1} < 1. \tag{8.5}$$

If the steady point $(1,0)$ is unstable, that is, if $\lambda_2 > 0$, then

$$\frac{\bar{m}}{\bar{a}+1} > 1 \tag{8.6}$$

and therefore $\bar{m} > 1$, so the second inequality in (8.5) is also satisfied. We conclude: If the equilibrium point $(1,0)$ is unstable then there exists another equilibrium point (c_*, b_*), where

$$c_* = \frac{\bar{a}}{\bar{m}-1}, \quad b_* = 1 - \frac{\bar{a}}{\bar{m}-1}. \tag{8.7}$$

The stability of the equilibrium point $(1,0)$ means that in steady state the chemostat does not produce any bacteria; this of course will not occur for chemostats which function well. Hence we shall now focus on the case where (8.6) holds.

Setting

$$\mu = \frac{\bar{a}}{\bar{m}-1}, \tag{8.8}$$

we can write the steady point (8.7) in the form $(\mu, 1 - \mu)$. The Jacobian at this point is

$$J(\mu, 1-\mu) = \begin{pmatrix} -1 - \frac{\bar{m}\bar{a}b}{(\bar{a}+c)^2} & -\frac{\bar{m}c}{\bar{a}+c} \\ \frac{\bar{m}\bar{a}b}{(\bar{a}+c)^2} & 0 \end{pmatrix}_{(\mu, 1-\mu)}.$$

Since trace $J(\mu, 1-\mu)$ is negative and $\det J(\mu, 1-\mu)$ is positive, the equilibrium point $(\mu, 1-\mu)$ is stable.

We can use the model to determine how to adjust some chemostat parameters in order to achieve the best throughput of bacteria. For example, suppose all the parameters of the chemostat are fixed, including C_0, but not the dilution rate D. We then want to determine how to best choose D in order to maximize the bacterial output.

But before doing that, let us see what the stability condition (8.6) means in terms of the dilution rate. We compute μ in terms of D (recall $\bar{a} = \frac{a}{C_0}$ and $\bar{m} = \frac{m}{D} = \frac{m_0}{VD}$),

$$\mu = \frac{\bar{a}}{\bar{m}-1} = \frac{aVD}{C_0(m_0 - VD)}.$$

Similarly,

$$\frac{\bar{m}}{\bar{a}+1} = \frac{m_0}{V}\frac{1/D}{a/C_0+1} = \frac{D_0}{D},$$

where

$$D_0 = \frac{m_0}{V}\frac{1}{a/C_0+1}. \tag{8.9}$$

Hence (8.6) holds and $(\mu, 1-\mu)$ is a stable steady equilibrium only if

$$D < D_0. \tag{8.10}$$

Thus, in order to avoid washout of the bacteria in the chamber of the chemostat, the dilution rate D should remain below D_0; if $D > D_0$ then there will be a washout.

We next observe that, since $\bar{t} = Dt$,

$$\frac{db}{dt} = \frac{db}{d\bar{t}}\frac{d\bar{t}}{dt} = D\frac{db}{d\bar{t}}, \quad \text{where } b = \frac{B}{\gamma C_0}.$$

Thus the effluent of b in steady state of the system (8.3)–(8.4) is D times the steady state component of b, $1-\mu$; the corresponding effluent of B is then $\gamma C_0 D$ times $1-\mu$. Writing, for $D < D_0$,

$$D(1-\mu) = D(1 - \frac{\bar{a}}{\bar{m}-1}) = D(1 - \frac{aVD}{C_0(m_0 - VD)}),$$

we conclude that to maximize the bacterial harvest one should take the dilution parameter D such that it maximizes the function

$$f(D) = D(1 - \frac{aVD}{C_0(m_0 - VD)}), \quad 0 < D < D_0. \tag{8.11}$$

We use standard Calculus to determine where the maximum is achieved. By direct computation we find that

$$f'(D) = \frac{g(D)}{(m_0 - VD)^2},$$

where

$$g(D) = \alpha D^2 + \beta D + m_0^2$$

and

$$\alpha = V^2(1 + \frac{a}{C_0}), \quad \beta = -2m_0 V(1 + \frac{a}{C_0}).$$

The two zeros of the polynomial $g(D)$, also the two zeros of $f'(D)$, are

$$D_{1,2} = \frac{1}{2\alpha}(-\beta \pm \sqrt{\beta^2 - 4\alpha m_0}),$$

where $0 < D_1 < D_2$, or in terms of the chemostat parameters,

$$D_{1,2} = \frac{1}{V^2(1+\frac{a}{C_0})}\left(m_0 V(1+\frac{a}{C_0}) \pm m_0 V\sqrt{(1+\frac{a}{C_0})\frac{a}{C_0}}\right).$$

By drawing the parabola $y = g(D)$ as in Fig. 8.1, we see that $g(D) < 0$ if $D_1 < D < D_2$ and $g'(D_1) < 0, g'(D_2) > 0$.

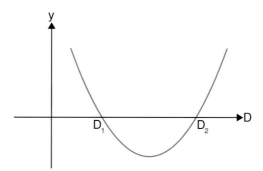

Fig. 8.1: The parabola $y = g(D)$.

By direct computation we also find that

$$g(D_0) = g(\frac{m_0/V}{1+\frac{a}{C_0}}) = \frac{m_0^2}{1+\frac{a}{C_0}} - m_0^2 < 0$$

and therefore

$$D_1 < D_0 < D_2.$$

Finally, since $g(D_1) = 0$,

$$f''(D_1) = \frac{g'(D_1)}{(m_0 - VD_1)^2} < 0.$$

Hence D_1 is the unique point where the maximum of $f(D)$ is achieved. We conclude:
In order to maximize the bacterial harvest the dilution rate should be chosen to be D_1.

Instead of attempting to determine the optimal choice of D by trial and error, our mathematical model provides an immediate and precise answer. Thus this example demonstrates the effectiveness of mathematical models.

The steady point $(\mu, 1 - \mu)$ is not only a local stable steady point but also a global stable point, that is, if

$$(c(0), b(0)) \neq (1, 0),$$

then

$$c(t) \to \mu, \quad b(t) \to 1 - \mu \quad \text{as } t \to \infty.$$

To prove it we set $z = c + b$ and add the two equations (8.3) and (8.4); we get

$$\frac{dz}{dt} + z = 1.$$

Hence $z(t) \to 1$ as $t \to \infty$. For simplicity we assume that $z(0) = 1$, and then $z(t) = 1$ for all $t > 0$. Substituting $c = 1 - b$ into Eq. (8.4) we find that the function $x = b(t)$ satisfies the equation

$$\frac{dx}{dt} = x\left(\frac{\bar{m}(1 - x)}{\bar{a} + (1 - x)} - 1\right) \equiv h(x).$$

We next check that

$$h'(x) > 0 \quad \text{if } x < 1 - \mu,$$
$$h'(x) < 0 \quad \text{if } x > 1 - \mu.$$

Hence if $x(0) < 1 - \mu$ then $x(t)$ increases to $1 - \mu$, and if $x(0) > 1 - \mu$ then $x(t)$ decreases to $1 - \mu$.

If initially $x(0) + z(0) \neq 1$ then

$$\frac{dx}{dt} = h(x, t), \quad \text{where } h(x, t) - h(x) \to 0 \text{ as } t \to \infty$$

since $z(t) \to 1$ as $t \to \infty$. One can deduce that, for any small $\varepsilon > 0$, if $x(0) < 1 - \mu - \varepsilon$ ($x(0) > 1 - \mu + \varepsilon$) then $x(t)$ increases (decreases) for all t large as long as $x(t)$ remains smaller than $1 - \mu - \varepsilon$ ($> 1 - \mu + \varepsilon$). Hence $x(t)$ enters the interval $(1 - \mu - \varepsilon, 1 - \mu + \varepsilon)$ after some time $t = \bar{t}$, and it will converge to $1 - \mu$ as $t \to \infty$ because $(\mu, 1 - \mu)$ is a stable point.

Problem 8.1. Consider a chemostat model given by the equations

$$\frac{dC}{dt} = \beta - CB - C,$$
$$\frac{dB}{dt} = CB - B,$$

where the inflow-outflow rate is 1. Show that

(i) if $\beta < 1$ then $(1, 0)$ is a stable steady point;

(ii) if $\beta > 1$ then $(1, 0)$ is an unstable steady point, and $(1, \beta - 1)$ is a stable steady point. This shows that in order to avoid washout, the rate β of nutrients supply must be larger than the flow rate.

Problem 8.2. Consider a chemostat with nutrients concentration z, and two types of bacteria, x and y:

$$\frac{dz}{dt} = \beta - zx - \mu zy - z, \quad \beta > 1,$$

$$\frac{dx}{dt} = zx - x,$$

$$\frac{dy}{dt} = \mu zy - y;$$

we take $\mu > 1$, which means that y is more efficient than x in consumption and growth. There are three steady points (z, x, y) where either x or y (or both) are washed out: $(1, \beta - 1, 0)$, $(\frac{1}{\mu}, 0, \beta - \frac{1}{\mu})$ and $(\beta, 0, 0)$. Determine which of these points are stable and give a biological interpretation.

Problem 8.3. Consider a chemostat model with two types of nutrients, x and y, consumed by bacteria z at different rates:

$$\frac{dx}{dt} = \beta - zx - x,$$

$$\frac{dy}{dt} = \beta - \mu zy - y,$$

$$\frac{dz}{dt} = zx + \mu zy - z.$$

Determine whether the steady point (x, y, z) given by

$$x = \frac{\beta}{1+z}, \quad y = \frac{\beta}{1+\mu z}, \quad x + \mu y = 1$$

is stable. [Hint: Use the Routh-Hurwitz criterion.]

8.1 Numerical Simulations

Problem 8.4. Consider the chemostat model of Problem 8.2 with $\beta = 3$, $\mu = 1.1$ so that bacteria y are more efficient than x in consumption and growth. Take $z(0) = 4$, $x(0) = 2$, $y(0) = \sigma < 2$ so that $x(t) > y(t)$ for t small. Simulate the model of Problem 8.2 for $t > 0$ until you arrive at time $t = T = T(\sigma)$ such that $y(T) = x(T)$; do it for $\sigma = 1, 1.1, 1.2, 1.3, 1.4, 1.5, 1.6, 1.7, 1.8, 1.9$, and draw the approximate curve $T = T(\sigma)$ for $0 < \sigma < 2$.

Problem 8.5. In Problem 8.4 take $\beta = 3$, $\mu = 1.1$ and $z(0) = 1$, $x(0) = 2$, $y(0) = \rho < 2$. Then nutrient x is consumed faster than nutrient y, but $x(t) > y(t)$ for t small. Find the fist time $t = T = T(\rho)$ such that $y(T) = x(T)$; do it for $\rho = 1, 1.1, 1.2, 1.3, 1.4, 1.5, 1.6, 1.7, 1.8, 1.9$, and draw the approximate curve $T = T(\rho)$ for $1 < \rho < 2$.

8.1.1 Finding the Steady States Using MATLAB

In the previous chapter, we have introduced basic schemes to calculate a root of an equation or a system. Here we explain how this can be done with MATLAB. There are several built-in functions in MATLAB that can be used to solve $\mathbf{f}(\mathbf{x}) = \mathbf{0}$:

```
>> x = fzero(fun,x0)
```

which attempts to find a zero of 'fun' near x_0 if x_0 is a scalar, or within an interval if x_0 is a vector, and 'fun' is a function handle. For example,

```
>> x = fzero(@cos,[1 2]) % cos(1) and cos(2) differ in
   sign
x = 1.5708
```

Another MATLAB function is 'fsolve':

```
>> x = fsolve(fun,x0)
```

It takes the initial guess x_0 and tries to solve the equations described in 'fun.' For example, suppose we would like to solve the following system:

$$F\begin{pmatrix} x_1 \\ x_2 \end{pmatrix} = \begin{pmatrix} 2x_1 - x_2 \\ -x_1 + 2x_2 \end{pmatrix} = \begin{pmatrix} e^{-x_1} \\ e^{-x_2} \end{pmatrix}$$

with the initial guess $(x_1, x_2) = (-5, -5)$. First, let's write a file that computes the values of F at a point $\mathbf{x} = (x_1, x_2)$, shown in Algorithm 8.1:

Algorithm 8.1. Function file for fsolve (myfun.m)

```
function F = myfun(x)
F = [2*x(1) - x(2) - exp(-x(1)); -x(1) + 2*x(2) - exp(-x(2))];
```

Next, we set up the initial point and options, and call fsolve (you can put the following commands in a file too):

```
>>x0 = [-5; -5]; % make an initial guess for the
        solution
>>options=optimset('Display','iter'); % option to
        display output
>> [x,fval] = fsolve(@myfun,x0,options) % call solver
```

After several iterations, 'fsolve' finds an answer as shown in Fig. 8.2. The iterations terminate because the vector of function values is near zero as measured by the default value of the function tolerance, and the problem appears regular as measured by the gradient. On the screen we can see the approximate solution x and its corresponding function value fval.

Iteration	Func-count	f(x)	Norm of step	First-order optimality	Trust-region radius
0	3	47071.2		2.29e+04	1
1	6	12003.4	1	5.75e+03	1
2	9	3147.02	1	1.47e+03	1
3	12	854.452	1	388	1
4	15	239.527	1	107	1
5	18	67.0412	1	30.8	1
6	21	16.7042	1	9.05	1
7	24	2.42788	1	2.26	1
8	27	0.032658	0.759511	0.206	2.5
9	30	7.03149e-06	0.111927	0.00294	2.5
10	33	3.29525e-13	0.00169132	6.36e-07	2.5

Fig. 8.2: Output for fsolve.

```
X =
   0.5671
   0.5671

fval =
1.0e-006 *
  -0.4059
  -0.4059
```

Problem 8.6. Use 'fsolve' to solve

$$x_1^3 + x_2 = 2,$$
$$x_2^3 - x_1 = -1.$$

Take the initial guess $(x_1, x_2) = (0, -5)$ and indicate how many steps it requires to reach the default tolerance for the solution. Write down the approximate solution.

Problem 8.7. Consider the model in (8.3)–(8.4). Take $a = 1, \bar{m} = 3$. Use 'fsolve' to numerically compute the steady state(s). Start with the initial guesses $(0,0)$ and $(5,5)$. Are the answers the same as the analytic solutions? Use the numerical solutions and the Jacobian to determine the local stability of the equilibrium points. [Hint: use 'eig' command.]

Chapter 9
Spread of Disease

Epidemiology is the study of patterns, causes, and effects of health and disease conditions in a population. It provides critical support for public health by identifying risk factors for disease and targets for preventive medicine. Epidemiology has helped develop methodology used in clinical research and public health studies. Major areas of epidemiological study include disease etiology, disease break, disease surveillance, and comparison of treatment effects such as in clinical trials.

Epidemiologists used gathered data and a broad range of biomedical and psychosocial theories to generate theory, test hypotheses, and make educated, informed assertions as to which relationships are causal and in which way. For example, many epidemiological studies are aimed at revealing unbiased relationships between exposure to smoking, biological agents, stress, or chemicals to mortality and morbidity. In the identification of causal relationship between these exposures and outcome epidemiologists use statistical and mathematical tools.

In this chapter we focus on epidemiology of infectious diseases. The adjectives **epidemic** and **endemic** are used to distinguish between a disease spread by an infective agent (epidemic) and a disease which resides in a population (endemic). For example, there are occasional spreads of the **cholera** epidemic in some countries, while **malaria** is endemic in Southern Africa. In this chapter we shall use mathematics in order to determine which epidemic will die out and which will become endemic.

In what follows we shall develop several different mathematical models for infectious diseases.

We begin with a simple model of a disease in a population of size N. We divide the population into three classes: susceptible S, infected I, and recovered R. Let

Electronic supplementary material The online version of this chapter (doi: 10.1007/978-3-319-29638-8_9) contains supplementary material, which is available to authorized users.

© Springer International Publishing Switzerland 2016
C.-S. Chou, A. Friedman, *Introduction to Mathematical Biology*,
Springer Undergraduate Texts in Mathematics and Technology,
DOI 10.1007/978-3-319-29638-8_9

β = infection rate,
μ = death rate, the same for all individuals,
v = recovery rate,
γ = rate by which recovered individuals have lost
their immunity and became susceptible to the disease.

Then we have the following diagram:

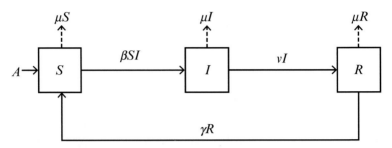

where A is the growth of susceptible. If all newborns are healthy, then not only S and R, but also I contributes to the growth term A. We view each of the populations S, I, R, N as representing a number of individuals (or a number density, that is, the number of individuals per unit area). The dimension of γ, μ, v is 1/time, the dimension of β is 1/(individual·time), and the dimension of A is individual/time. Based on the above diagram, we set up the following system of differential equations:

$$\frac{dS}{dt} = A - \beta SI + \gamma R - \mu S,$$
$$\frac{dI}{dt} = \beta SI - vI - \mu I, \qquad (9.1)$$
$$\frac{dR}{dt} = vI - \gamma R - \mu R.$$

To examine more carefully the meaning of A, we introduce a differential equation for $N(t)$, which is obtained by adding all the equations in (9.1),

$$\frac{dN}{dt} = A - \mu N.$$

Given an initial population density N_0, we find that

$$N(t) = N_0 e^{-\mu t} + \frac{A}{\mu}(1 - e^{-\mu t}).$$

Hence $N(t) \to A/\mu$ as $t \to \infty$. Thus A/μ is equal to the asymptotic density of the population (as $t \to \infty$).

The system (9.1) is called the **SIR model**. The SIR model has an equilibrium point which is disease free, namely

$$(S_0, I_0, R_0) = (\frac{A}{\mu}, 0, 0);$$

we call it the **disease free equilibrium** (DFE). The Jacobian matrix at the DFE is

$$\begin{pmatrix} -\mu & -\beta\frac{A}{\mu} & \gamma \\ 0 & \beta\frac{A}{\mu} - (v+\mu) & 0 \\ 0 & v & -\mu - \gamma \end{pmatrix}.$$

The characteristic polynomial is

$$(\mu + \lambda)(\beta\frac{A}{\mu} - (v+\mu) - \lambda)(\mu + \gamma + \lambda),$$

and the eigenvalues are $\lambda_1 = -\mu, \lambda_2 = -\mu - \gamma, \lambda_3 = \beta\frac{A}{\mu} - (v+\mu)$. Hence the DFE is stable if

$$\beta\frac{A}{\mu} < v + \mu. \tag{9.2}$$

We conclude that in order to stop the spread of infection we need to either decrease the rate of infection (β) by decreasing contact between healthy and infected individuals, or by increasing the rate of recovery (v) by drugs. When (9.2) holds, any new small infection will die out with time. On the other hand if

$$\beta\frac{A}{\mu} > v + \mu, \tag{9.3}$$

the DFE is unstable; there are arbitrarily small infections that will not disappear in the population. Furthermore, there is an equilibrium point $(\bar{S}, \bar{I}, \bar{R})$ with $\bar{I} > 0$, namely

$$\beta\bar{S} = v + \mu, \quad \bar{R} = \frac{v}{\gamma + \mu}\bar{I}, \quad \frac{\beta}{\mu}\bar{I} = \frac{(\beta\frac{A}{\mu} - (v+\mu))}{v + \mu - \frac{\gamma v}{\gamma + \mu}}. \tag{9.4}$$

Note that $v + \mu - \gamma v/(\gamma + \mu) > 0$, and hence the inequality (9.3) ensures that $\bar{I}$ is positive.

It is natural to ask whether the equilibrium point $(\bar{S}, \bar{I}, \bar{R})$ is stable. The answer is yes, as asserted by the following theorem.

Theorem 9.1. *If (9.3) holds then the equilibrium point $(\bar{S}, \bar{I}, \bar{R})$ is stable.*

Proof. The Jacobian matrix at $(\bar{S}, \bar{I}, \bar{R})$ is

$$J = \begin{pmatrix} -\beta\bar{I} - \mu & -\beta\bar{S} & \gamma \\ \beta\bar{I} & \beta\bar{S} - (v+\mu) & 0 \\ 0 & v & -(\gamma + \mu) \end{pmatrix}$$

and $\beta\bar{S} = v + \mu$. Hence the characteristic polynomial is

$$\det(J - \lambda I) = \begin{vmatrix} -\beta\bar{I} - \mu - \lambda & -(v+\mu) & \gamma \\ \beta\bar{I} & -\lambda & 0 \\ 0 & v & -(\gamma + \mu) - \lambda \end{vmatrix}.$$

Expanding the determinant by the first column we derive the characteristic equation

$$(\beta\bar{I}+\mu+\lambda)[\lambda^2+(\gamma+\mu)\lambda]+\beta\bar{I}[(v+\mu)(\gamma+\mu+\lambda)-v\gamma]=0,$$

or

$$\lambda^3+\alpha_1\lambda^2+\alpha_2\lambda+\alpha_3=0,$$

where

$$\alpha_1=(\beta\bar{I}+\mu)+(\gamma+\mu),$$
$$\alpha_2=(\beta\bar{I}+\mu)(\gamma+\mu)+\beta\bar{I}(v+\mu),$$
$$\alpha_3=\beta\bar{I}[(v+\mu)(\gamma+\mu)-v\gamma].$$

Clearly all the α_i are positive, and

$$\alpha_1\alpha_2=\beta\bar{I}(v+\mu)(\gamma+\mu)+\text{positive terms}>\beta\bar{I}[(v+\mu)(\gamma+\mu)-v\gamma]=\alpha_3.$$

Hence, by the Routh-Hurwitz criterion, the equilibrium point $(\bar{S},\bar{I},\bar{R})$ is stable.

A stable equilibrium point with $I>0$ is called **endemic**; it represents a disease that will never disappear.

An important concept in epidemiology is the basic reproduction number, defined as follows: In a healthy population we introduce one infection and compute the expected infection among the susceptibles caused by this single infection. We call it the **expected secondary infection**, or **basic reproduction number**, and denote it by R_0. Then intuitively it is clear that DFE is stable if $R_0<1$ (the secondary infection is smaller than the initial infection) whereas if $R_0>1$ then the DFE will be unstable.

Consider, for example, the SIR model (9.1). The DFE is $(A/\mu,0,0)$. One infection evolves according to

$$\frac{dI}{dt}=-vI-\mu I,\quad I(0)=1,$$

so that $I(t)=e^{-(v+\mu)t}$ at time t, with total life-time infection

$$\int_0^\infty I(t)dt=\frac{1}{v+\mu}.$$

The secondary infection in healthy population is then

$$R_0=\beta\frac{A}{\mu}\frac{1}{v+\mu}.$$

As already computed in (9.2) and (9.3), the DFE is stable if $R_0<1$ and unstable if $R_0>1$.

When a susceptible is exposed to an infected individual, he/she may or may not become immediately sick. With this in mind, we may extend the SIR model by introducing a new class E of exposed individuals. The new model, called the **SEIR model**, consists of the following equations:

$$
\begin{aligned}
\frac{dS}{dt} &= A - \beta SI + \gamma R - \mu S,\\
\frac{dE}{dt} &= \beta SI - \kappa E - \mu E,\\
\frac{dI}{dt} &= \kappa E - \nu I - \mu I,\\
\frac{dR}{dt} &= \nu I - \gamma R - \mu R.
\end{aligned}
\tag{9.5}
$$

Here κ is the rate by which the exposed become infected, and β is the rate of infection of susceptibles by infected individuals. The DFE for the SEIR model is $(\frac{A}{\mu},0,0,0)$.

Problem 9.1. Show that the DFE of the system (9.5) is stable if

$$
\beta \frac{A}{\mu} < \frac{(\nu + \mu)(\kappa + \mu)}{\kappa}.
$$

Problem 9.2. Prove that if the DFE of the system (9.5) is not stable, then there exists another equilibrium point.

In the SIR model we have taken the infection term to be βSI, that is, it depends on the **density** of the infected individuals. Another possibility is to take the infection term to be $\frac{\beta SI}{N}$, where $\frac{I}{N}$ is the relative proportion of the infected individuals, namely, the **frequency** or **prevalence** of the infection.

Problem 9.3. Show that when βSI is replaced by $\frac{\beta SI}{N}$ in (9.1), where $N = S + I + R$, the DFE $(\frac{A}{\mu},0,0)$ is stable if $\beta < \nu + \mu$.

Problem 9.4. If in the previous problem (9.3) is replaced by $\beta > \nu + \mu$, then the DFE is not stable, and there exists another equilibrium point.

So far we considered infectious diseases that do not cause death. In infectious diseases which cause death, e.g., Ebola, we need to change the equation for I by including a death rate ρ caused by the disease. Then the equation for I in the system (9.1) becomes

$$
\frac{dI}{dt} = \beta SI - \gamma I - \mu I - \rho I.
\tag{9.6}
$$

Problem 9.5. Show that if the equation for I in the system (9.1) is changed into (9.6), then the disease-free equilibrium $(\frac{A}{\mu},0,0)$ is stable if

$$
\beta \frac{A}{\mu} < \nu + \mu + \rho.
$$

Problem 9.6. Prove that if in the system (9.1), with the equation for I changed into (9.6),

$$\beta \frac{A}{\mu} > v + \mu + \rho,$$

then there exists an equilibrium point $(\bar{S}, \bar{I}, \bar{R})$, and it is stable.

9.1 HIV

In humans infected with HIV, the HIV virus enters the CD4$^+$ T cells and hijack the machinery of the cells in order to multiply within these cells. As an infected T cell dies, an increased number of virus particles emerge to invade and infect new CD4$^+$ T cells. This process eventually leads to significant depletion of the CD4$^+$ T cells, from over 700 in mm^3 of blood in healthy individuals to 200 in mm^3. This state of the disease is characterized as AIDS; the immune system is too weak to sustain life for too long. In order to determine whether an initial infection with HIV will develop into AIDS we introduce a simple model which includes the CD4$^+$ T cells, denoted by T, the infected CD4$^+$ T cells, denoted by T^*, and the HIV virus outside the T cells, denoted by V. Their number densities satisfy the following system of equations:

$$\frac{dT}{dt} = A - \beta TV - \mu T,$$
$$\frac{dT^*}{dt} = \beta TV - \mu^* T^*, \qquad\qquad (9.7)$$
$$\frac{dV}{dt} = \gamma \mu^* T^* - \kappa V.$$

Here A is the natural production rate of healthy T cells, β is the infection rate of healthy T cells by external virus, μ and μ^* are the death rates of T and T^*, respectively, γ is the number of virus particles that emerge upon death of infected one CD4$^+$ T cell, and κ is the death rate of the virus.

Problem 9.7. In the model (9.7), the DFE is $\left(\frac{A}{\mu}, 0, 0\right)$. Prove that the DFE is stable if

$$\frac{\beta A}{\mu} < \frac{\kappa}{\gamma},$$

and is unstable if this inequality is reversed.

We can compute the basic reproduction number R_0 for the model (9.7) as follows: One virion has the life time of $\frac{1}{\kappa}$ (since $\frac{dV}{dt} = -\kappa V$, $V(t) = e^{-\kappa t}$, $\int_0^\infty V(t)dt = \frac{1}{\kappa}$) and it infects (A/μ) T cells at rate β, whereas each infected T cells gives rise to γ virus particles. Hence the basic reproduction number is

$$R_0 = \frac{1}{\kappa}\beta\frac{A}{\mu}\gamma = \frac{\beta A \gamma}{\kappa \mu}.$$

From Problem 9.7 we see that the DFE is stable if $R_0 < 1$ and is unstable if $R_0 > 1$.

More information about how to calculate the basic reproduction number R_0 for infectious diseases is given in article [5].

9.2 Numerical Simulations

Problem 9.8. It seems reasonable to expect that the infected population $I(t)$ in the SIR model should increase, at each time t, if β is increased. Consider the special case where $A = 2$, $v = \gamma = \mu = 1$, in which case the DFE $(2,0,0)$ is stable if $\beta < 1$ and unstable if $\beta > 1$. Take initial values

$$(S(0), I(0), R(0)) = (1.8, 0.2, 0)$$

and simulate the curve $I(t)$ for $0 < t < 20$, with $\beta = 0.9, 0.95, 1, 1.05, 1.1$. Are these curves $I_\beta(t) \equiv I(t)$ increasing with β, for each t?

Algorithm 9.1. Main file for SIR model in Problem 9.8 (main_SIR.m)

```
%% SIR model for Problem 9.8
global A nu gamma mu beta

%% parameters
A = 2;
nu = 1;
gamma = 1;
mu = 1;
beta = 0.9;

%% initial conditions
S0 = 1.8; % susceptible
I0 = 0.2; % infected
R0 = 0;   % recovered

init = [S0; I0; R0];
tspan = [0,20];

[t,v] = ode45('fun_SIR',tspan,init);

plot(t,v(:,2)), hold on
```

Problem 9.9. HIV is an incurable disease so that, in the model (9.6),

$$\frac{\beta A}{\mu} > \frac{k}{\gamma}, \quad \text{or } \beta\gamma > \frac{\mu k}{A}.$$

Algorithm 9.2. fun_SIR.m

```
function dv = fun_SIR(t,v)
global A nu gamma mu beta

S = v(1); % susceptible
I = v(2); % infected
R = v(3); % recovered
dv = zeros(3,1);

dv(1) = A - beta*S*I + gamma*R - mu*S;
dv(2) = beta*S*I - nu*I - mu*I;
dv(3) = nu*I - gamma*R - mu*R;
```

Then there is a unique steady state $(\bar{T}, \bar{T}^*, \bar{V})$ with

$$\bar{T} = \frac{k}{\beta\gamma}, \quad \beta\bar{V} = \frac{A\beta\gamma}{k} - \mu, \quad \bar{T}^* = \frac{\beta}{\mu^*}\bar{T}\bar{V}.$$

Anti-HIV drug that blocks invasion of extracellular virus into T cells decreases the parameter β, and anti-HIV drug that reduces the replication of virus within T^* cells decreases the parameter γ. By either decreasing β or γ, or both, the steady state $\bar{T}$ is increased. Suppose initially the parameters β and γ were such that $\beta \geq 1, \gamma \geq 1$, and take $A = 500, k = 1000, \mu = 0.5, \mu^* = 1$. Consider three sets of parameters (but $\beta\gamma$ is still greater than $\frac{\mu k}{A}$): (i) $\beta = 4, \gamma = 4$; (ii) $\beta = 1, \gamma = 4$; (iii) $\beta = 1, \gamma = 2$.

Starting with $T(0) = 400, T^*(0) = 200, V(0) = 10$, simulate the system (9.7) for $0 \leq t \leq 60$ under these three parameter sets. What do you observe? Explain the results.

Chapter 10
Enzyme Dynamics

Cells are the basic units of life. A cell consists of a concentrated aqueous solution of molecules contained in a membrane, called **plasma membrane**. A cell is capable of replicating itself by growing and dividing. Cells that have a nucleus are called **eukaryotes**, and cells that do not have a nucleus are called **prokaryotes**. Bacteria are prokaryotes, while yeast and amoebas, as well as most cells in our body, are eukaryotes. The **Deoxyribonucleic acid (DNA)** are very long polymeric molecules, consisting of two strands of chains, having double helix configuration, with repeated nucleotide units A, C, G, and T. The DNA is packed in chromosomes, within the nucleus in eukaryotes. In humans, the number of chromosomes is 46, except in sperm and egg cells where the number is 23.

The DNA is the genetic code of the cell; it codes for proteins. Proteins lie mostly in the cytoplasm of the cells, that is, outside the nucleus; some proteins are attached to the plasma membrane, while some can be found in the nucleus. Proteins are polymers of amino acids whose number typically ranges from hundreds to thousands; there are 20 different amino acids from which all proteins are made. Each protein assumes 3-dimensional configuration, called **conformation**. Proteins perform specific tasks by changing their conformation.

Two proteins, A and B, may combine to form a new protein C. We express this process by writing

$$A + B \rightarrow C.$$

Biological processes within a cell involves many such reactions. Some of these reactions are very slow, others are very fast, and in some cases the reaction rate may start slow, then speed up until it reaches a maximal level. In this chapter we consider the question: How to determine the speed of biochemical reactions among proteins? In order to address this question we shall develop some mathematical models.

Electronic supplementary material The online version of this chapter (doi: 10.1007/978-3-319-29638-8_10) contains supplementary material, which is available to authorized users.

© Springer International Publishing Switzerland 2016
C.-S. Chou, A. Friedman, *Introduction to Mathematical Biology*,
Springer Undergraduate Texts in Mathematics and Technology,
DOI 10.1007/978-3-319-29638-8_10

We begin with a simple case. Suppose we have two proteins, A and B, or more generally, two molecules A and B. We assume that A and B, when coming in contact, undergo a reaction, at some rate k_1, that makes them form a new molecule C. We express this reaction by writing

$$A + B \xrightarrow{k_1} C;$$

k_1 is called the **rate coefficient**. We shall denote the respective concentrations of three molecules by $[A]$, $[B]$, and $[C]$. The **law of mass action** states that the reaction rate $\frac{d[C]}{dt}$, or v_1, of the above reaction is given by

$$v_1 = k_1[A][B],$$

that is,

$$\frac{d[C]}{dt} = k_1[A][B].$$

Note that the above reaction implies that

$$\frac{d[A]}{dt} = -k_1[A][B], \quad \frac{d[B]}{dt} = -k_1[A][B].$$

If the reaction is reversible with rate coefficient k_{-1}, then

$$A + B \underset{k_{-1}}{\overset{k_1}{\rightleftharpoons}} C$$

and

$$\frac{d[C]}{dt} = k_1[A][B] - k_{-1}[C],$$

$$\frac{d[A]}{dt} = \frac{d[B]}{dt} = -k_1[A][B] + k_{-1}[C].$$

The law of mass action can be extended to interaction among three or more molecules. Consider for example three species X_1, X_2, X_3 that interact to form a species Y:

$$X_1 + X_2 + X_3 \xrightarrow{k} Y$$

where k is the reaction rate. Then the law of mass action states that

$$\frac{d[X_i]}{dt} = -k[X_1][X_2][X_3] \quad \text{for } i = 1, 2, 3.$$

In particular, if $X_1 = A$, $X_2 = X_3 = B$, $Y = C$, then

$$A + 2B \xrightarrow{k} C$$

and

$$\frac{d[A]}{dt} = -k[A][B]^2,$$

$$\frac{d[B]}{dt} = -2k[A][B]^2,$$

$$\frac{d[C]}{dt} = k[A][B]^2.$$

Example 10.1. Consider the chemical reactions

$$A + B \xrightarrow{1} C, \quad C \xrightarrow{2} A + B$$

with initial concentrations

$$[A(0)] + [C(0)] = 2, \quad [B(0)] - [A(0)] = 1.$$

We wish to determine the behavior of the concentrations as $t \to \infty$. To do that we set

$$x(t) = [A(t)], \quad y(t) = [B(t)], \quad z(t) = [C(t)].$$

Then by the law of mass action,

$$\frac{dx}{dt} = -xy + 2z,$$

$$\frac{dy}{dt} = -xy + 2z,$$

$$\frac{dz}{dt} = xy - 2z.$$

Hence

$$\frac{d(x+z)}{dt} = 0, \quad \frac{d}{dt}(y+z) = 0, \quad \frac{d}{dt}(y-x) = 0$$

so that

$$x(t) + z(t) \equiv constant, \quad y(t) + z(t) \equiv constant, \quad y(t) - x(t) \equiv constant.$$

Recalling the initial conditions, we see that

$$y(t) = x(t) + 1, \quad z(t) = 2 - x(t).$$

We can then write the differential equation for $x(t)$ in the form

$$\frac{dx}{dt} = -x(1+x) + 2(2-x) = -x^2 - 3x + 4 = -(x+4)(x-1),$$

or

$$\frac{dx}{(x+4)(x-1)} = -dt.$$

By integration, using the relation

$$\frac{1}{(x+4)(x-1)} = \frac{1}{3}\left(\frac{1}{x-1} - \frac{1}{x+4}\right),$$

we obtain

$$\frac{1}{3}\ln\frac{|x-1|}{x+4} = -t + constant.$$

It follows that $x(t) \to 1$ as $t \to \infty$, and then also $y(t) \to 2$ and $z(t) \to 1$ as $t \to \infty$.

Example 10.2. Consider the chemical reactions

$$A + B \xrightarrow{k} C, \quad B + C \xrightarrow{k} A.$$

We want to determine how the concentrations of these chemicals will change as $t \to \infty$. Setting

$$x(t) = [A(t)], \quad y(t) = [B(t)], \quad z(t) = [C(t)]$$

we have, by the law of mass action:

$$\frac{dx}{dt} = -kxy + kyz,$$

$$\frac{dy}{dt} = -kxy - kyz,$$

$$\frac{dz}{dt} = kxy - kyz.$$

We observe that

$$\frac{d}{dt}(x+z) = 0, \quad \text{hence } x(t) + z(t) = \alpha,$$

where $\alpha = x(0) + z(0)$. Then the equation for y can be written in the form

$$\frac{dy}{dt} = -ky(x+z) = -k\alpha y.$$

If $y(0) = \beta$ then

$$y(t) = \beta e^{-k\alpha t}. \tag{10.1}$$

Hence $y(t) \to 0$ as $t \to \infty$. Next we write

$$\frac{dx}{dt} = -kxy + ky(\alpha - x) = -2kxy + k\alpha y,$$

or, by (10.1),

$$\frac{dx}{dt} + 2k\beta e^{-k\alpha t}x = k\alpha\beta e^{-k\alpha t}.$$

This is a linear differential equation of the type considered in Section 2.2.1. Introducing the integral

$$P(t) = \int_0^t 2k\beta e^{-kas}ds = -\frac{2\beta}{\alpha}(e^{-kat} - 1),$$

we can solve for $x(t)$ using the formula (2.4):

$$x(t) = e^{-P(t)}x(0) + k\alpha\beta e^{-P(t)}\int_0^t e^{P(s)}e^{-kas}ds.$$

To compute the last integral we substitute

$$z = e^{-kas}, \quad dz = -kae^{-kas}ds$$

Then the integral becomes

$$-\frac{e^{\frac{2\beta}{\alpha}}}{k\alpha}\int_{z(0)}^{z(t)}e^{-\frac{2\beta}{\alpha}z}dz = e^{\frac{2\beta}{\alpha}}\frac{e^{-\frac{2\beta}{\alpha}z}}{2k\beta}\bigg|_{z(0)}^{z(t)} = \frac{1}{2k\beta}\left(e^{\frac{-2\beta}{\alpha}(e^{-kat}-1)} - 1\right)$$

so that

$$x(t) = e^{\frac{2\beta}{\alpha}(e^{-kat}-1)}x(0) + \frac{\alpha}{2}\left(1 - e^{\frac{2\beta}{\alpha}(e^{-kat}-1)}\right).$$

At $t \to \infty$, $e^{-kat} \to 0$, and hence $x(t) \to e^{\frac{-2\beta}{\alpha}}x(0) + \frac{\alpha}{2}(1 - e^{-\frac{2\beta}{\alpha}})$, and $z(t) \to -e^{\frac{-2\beta}{\alpha}}x(0) + \frac{\alpha}{2}(1 + e^{-\frac{2\beta}{\alpha}})$.

Metabolism in a cell is the sum of physical and chemical processes by which material substances are produced, maintained, or destroyed, and by which energy is made available. **Enzymes** are proteins that act as catalysts in speeding up chemical reactions within a cell. They play critical roles in many metabolic processes within the cell. An enzyme, say E, can take a molecule S and convert it to a molecule P in one millionth of a second. The original molecule S is referred to as the **substrate**, and P is called the **product**. The enzyme-catalyzed conversion of a substrate S into a product P is written in the form

$$S \xrightarrow{E} P. \tag{10.2}$$

Figure 10.1 illustrates how an enzyme can convert substrate S into a product P.

The profile $[S] \to [P]$ can take different forms, depending on the underlying biology. Two typical profiles are shown in Figure 10.2.

Figures 10.2(A) and 10.2(B) have been shown to hold in different experiments, but it would be useful to derive them by mathematical analysis based on known properties of enzymes. Indeed such a derivation will give us a precise mathematical formula for the profiles displayed in Figure 10.2. We begin with the derivation of a formula that yields the profile of Figure 10.2(A).

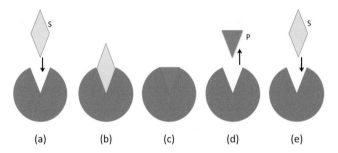

Fig. 10.1: (a) Enzyme attracts S; (b) S is inside E; (c) Enzymatic process converts S into P; (d) P is released; (e) Enzyme is ready to attract another S.

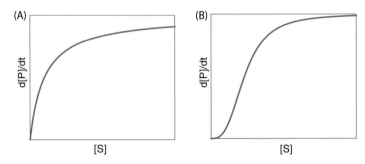

Fig. 10.2: Two different profiles of the enzymatic conversion of $S \rightarrow P$. (A) Michaelis-Menten kinetics; (B) Hill kinetics of order 3.

In the following we show how such a profile can be derived from the law of mass action. We write, schematically,

$$S + E \underset{k_{-1}}{\overset{k_1}{\rightleftharpoons}} C,$$

where C is the complex SE,

$$C \xrightarrow{k_2} E + P.$$

By the law of mass action

$$\frac{d[C]}{dt} = k_1[S][E] - (k_{-1} + k_2)[C], \qquad (10.3)$$

$$\frac{d[E]}{dt} = -k_1[S][E] + (k_{-1} + k_2)[C], \qquad (10.4)$$

$$\frac{d[S]}{dt} = -k_1[S][E] + k_{-1}[C], \qquad (10.5)$$

$$\frac{d[P]}{dt} = k_2[C]. \qquad (10.6)$$

Notice that

$$\frac{d}{dt}([E]+[C]) = 0$$

so that $[E]+[C] = constant = e_0$; e_0 is the total concentration of the enzyme in both E and the complex C. Note that $\frac{d[C]}{dt} + \frac{d[S]}{dt} + \frac{d[P]}{dt} = 0$, so Eq. (10.5) depends on Eqs. (10.3) and (10.6) and may therefore be dropped.

We focus on Eq. (10.3) and note that in the enzymatic process the complex C changes very fast, so that compared to the other equations, (10.4)–(10.6), $d[C]/dt$ is approximately zero. Hence, we approximate Eq. (10.3) by the steady state equation

$$k_1[S][E] - (k_{-1}+k_2)[C] = 0.$$

Substituting $[E] = e_0 - [C]$ we get

$$k_1[S](e_0 - [C]) = (k_{-1}+k_2)[C],$$

or

$$[C] = \frac{k_1 e_0 [S]}{(k_{-1}+k_2) + k_1[S]} = \frac{e_0[S]}{k_M + [S]},$$

where $K_M = \frac{k_{-1}+k_2}{k_1}$.

Then

$$\frac{d[P]}{dt} = k_2[C] = \frac{V_{max}[S]}{K_M + [S]}, \tag{10.7}$$

where $V_{max} = k_2 e_0$.

We have thus derived the **Michaelis-Menten formula**

$$\frac{d[P]}{dt} = \frac{V_{max}[S]}{K_M + [S]}, \tag{10.8}$$

where V_{max} and K_M are constants; note that

$$\frac{d[P]}{dt} \to V_{max} \quad \text{as } [S] \to \infty.$$

The assumption we made in the derivation of (10.8) that $d[C]/dt$ is very small is quite reasonable and, indeed, the Michaelis-Menten formula is widely used in describing enzymatic processes.

But what about Figure 10.2(B)? Such a profile is based on a different enzymatic process, for example when an enzyme E can bind first to one substrate S and then with another substrate S. Furthermore, in such a case, as is well established experimentally, the speed by which the enzyme binds to the second substrate is much faster, as illustrated in Figure 10.3.

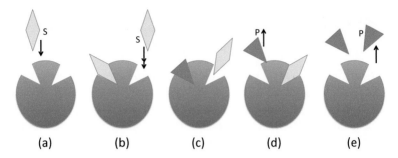

Fig. 10.3: Enzyme with two sites for absorbing and converting substrate S to product P; the conversion of the second substrate is faster than the conversion of the first substrate.

We model such processes as follows:

$$\begin{aligned}
S + E &\underset{k_{-1}}{\overset{k_1}{\rightleftharpoons}} C_1, &&(C_1 = SE)\\
C_1 &\overset{k_2}{\rightarrow} E + P,\\
S + C_1 &\underset{k_{-3}}{\overset{k_3}{\rightleftharpoons}} C_2, &&(C_2 = SC_1 = S^2 E)\\
C_2 &\overset{k_4}{\rightarrow} C_1 + P,
\end{aligned}$$

(10.9)

so that

$$\frac{d[P]}{dt} = k_2[C_1] + k_4[C_2].$$

Note that $[E] + [C_1] + [C_2] = constant = e_0$. Assuming the steady state approximations

$$\frac{d[C_1]}{dt} = \frac{d[C_2]}{dt} = 0,$$

one can show that

$$\frac{d[P]}{dt} = \frac{(k_2 K_2 + k_4[S])e_0[S]}{K_1 K_2 + K_2[S] + [S]^2},$$

(10.10)

where

$$K_1 = \frac{k_{-1} + k_2}{k_1}, \quad K_2 = \frac{k_{-3} + k_4}{k_3}.$$

Steps 1 and 3 in equations of (10.9) represent sequential binding of two substrate molecules to the enzyme. We assume that previously enzyme-bound substrate molecule significantly increases the rate of binding of a second substrate molecule, so that $k_3 \gg k_1$. In the extreme case of $k_1 \to 0$, $k_3 \to \infty$, with $k_1 k_3$ a finite positive constant, we get $K_1 \to \infty$, $K_2 \to 0$, $K_1 K_2 \to K_H > 0$, so that

$$\frac{d[P]}{dt} = \frac{V_{max}[S]^2}{K_H + [S]^2},$$ (10.11)

where V_{max} and K_H are constants. Formula (10.11) is called the **Hill kinetics**.

Some enzymes can bind to three or more substrates. In this case it is often the case that when enzyme has already bound to m substrates S, it has a greater affinity to bind to the next substrate S. Under this biological assumption, one can derive the Hill kinetics of order n,

$$\frac{d[P]}{dt} = \frac{V_{max}[S]^n}{K_H + [S]^n}.$$ (10.12)

Figure 10.2(B) displays a profile of Hill kinetics of order 3. The Michaelis-Menten formula is used also in other biological processes. For example, when macrophages M ingest bacteria B they become infected macrophages M_i. The resulting growth in M_i is described by the Michaelis-Menten formula

$$\frac{d[M_i]}{dt} = \lambda[M]\frac{[B]}{K + [B]}.$$

Notice that for small $[B]$, this is approximately the law of mass action of

$$M + B \rightarrow M_i.$$

However the capacity of macrophages to ingest bacteria is limited by the following fact: After receptor proteins on the macrophage membrane have been engaged in the ingestion process, they need to take time off for recycling. Hence there is a limit, λ, on how fast macrophages can ingest the bacteria.

Problem 10.1. Consider the chemical reaction

$$A + 2B \xrightarrow{k} C,$$

where initially $2[A(0)] - [B(0)] = 1$. Show that $y(t) = [B(t)]$ satisfies the equation

$$y' = -ky^2(1 + y),$$

that the solution of the above equation is given by

$$-\frac{1}{y} + \ln\frac{1 + y}{y} = -kt + C, \quad C \text{ is a constant,}$$

and that $y(t) \rightarrow 0$ as $t \rightarrow \infty$

Problem 10.2. Consider the chemical reactions

$$A + B \xrightarrow{k} C, \quad B + C \xrightarrow{2k} A$$

with initial condition $[A(0)] + [C(0)] = 1.$ Prove that $y(t) = [B(t)]$ satisfies the equation

$$\frac{dy}{dt} = (kx - 2k)y,$$

and the inequality $y(t) \le e^{-kt} y(0)$.

Problem 10.3. Derive Eq. (10.10) under the steady state approximations $d[C_1]/dt = 0$, $d[C_2]/dt = 0$.

10.1 Numerical Simulations

Problem 10.4. Suppose

$$A + B \xrightarrow{k} C, \quad C \xrightarrow{3} A + B$$

Set $x = [A]$, $y = [B]$, $z = [C]$ and take $x(0) = y(0) = 1$, $z(0) = 8$. Derive a system of differential equations for $x(t), y(t), z(t)$, and compute $x(10)$ as a function of k, for $1 \le k \le 5$. Sample codes are shown in Algorithms 10.1 and 10.2.

Problem 10.5. Suppose

$$A + B \xrightarrow{k} C, \quad C \xrightarrow{3} A + 2B$$

Set $x = [A]$, $y = [B]$, $z = [C]$ and take $x(0) = y(0) = 1$, $z(0) = 8$. Derive a system of differential equations for $x(t), y(t), z(t)$, and compute $x(10)$ as a function of k, for $1 \le k \le 5$.

Algorithm 10.1. Main file for Problem 10.4 (main_EnzymeDynamics.m)

```
%%% This code is a sample code for Problem 10.4
global k

Tfinal = 10;
tspan = [0,Tfinal];

%% set up initial conditions
A_0 = 1;
B_0 = 1;
C_0 = 8;

z_ini = [A_0; B_0; C_0];

%% discretize k in [1,5]
n = 51;  % 51 discretization points including 1 and 5
K = linspace(1,5,n);

x_10 = zeros(n,1); % define a vector to store x(10) for each k

%% for each value in vector K, solve the ODE
for i = 1 : n
    k = K(i);
    [t,z] = ode45('fun_EnzymeDynamics',tspan,z_ini);
    x_10(i) = z(end,1);
end

%% plot
plot(K,x_10)
xlabel('k'); ylabel('x(10)');
title(['T = ' num2str(Tfinal)]);
```

Algorithm 10.2. fun_EnzymeDynamics.m

```
%%% This code is a function file for the main code for Problem
  % 10.4
function dz = fun_EnzymeDynamics(t,z)
global k

dz = zeros(3,1);
A  = z(1);
B  = z(2);
C  = z(3);

dz(1) = - k*A*B + 3*C;
dz(2) = - k*A*B + 3*C;
dz(3) = k*A*B - 3*C;
```

Chapter 11
Bifurcation Theory

Consider two populations, x and y, that are interacting either by competition, or as predator and prey. They may end up near a stable steady state, or possibly in seasonally oscillating states; this could depend on their proliferation rates, death rates, available resources, climate change, etc. In this chapter we wish to explore these varied possibilities using mathematics. To do that we begin by a short introduction to the theory of bifurcations. **Bifurcation theory** is concerned with the question of how the behavior of a system which depends on a parameter p changes with the parameter. It focuses on any critical value, $p = p_{cr}$, where the behavior of the system undergoes radical change; such values are called **bifurcation points**. The change that occurs at $p = p_c$ typically involves two or more branches of solutions which depend on the parameter p; the nature of these 'bifurcation' branches changes radically at $p = p_c$.

We shall consider bifurcation phenomena for a system of differential equations with parameter p,

$$\frac{d\mathbf{x}}{dt} = \mathbf{f}(\mathbf{x}, p). \tag{11.1}$$

Bifurcation points can arise in different ways. For example, suppose a steady state of Eq. (11.1), which depends on p, is stable for $p < p_c$ but loses stability at p_c. Then a qualitative change has occurred in the phase portrait of the system (11.1), and $p = p_c$ is a bifurcation point. It sometimes happens that as p increases from $p < p_c$ to $p > p_c$ the differential system will begin to have periodic solutions, a well-recognized biological phenomena. Thus we would like to determine, mathematically, when such a situation takes place.

Problems 11.1–11.3 are simple but typical examples of bifurcations that frequently occur in biology.

Electronic supplementary material The online version of this chapter (doi: 10.1007/978-3-319-29638-8_11) contains supplementary material, which is available to authorized users.

C.-S. Chou, A. Friedman, *Introduction to Mathematical Biology*,
Springer Undergraduate Texts in Mathematics and Technology,
DOI 10.1007/978-3-319-29638-8_11

Problem 11.1. Consider the equation

$$\frac{dx}{dt} = p + x^2.$$

It has two steady states $x = \pm\sqrt{-p}$ if $p < 0$ and no steady states if $p > 0$. Prove that $x = -\sqrt{-p}$ is stable and $x = +\sqrt{-p}$ is unstable. The point $p = 0$ is called a **saddle-point** bifurcation.

Problem 11.2. Consider the equation

$$\frac{dx}{dt} = px - x^2.$$

It has steady points $x = 0$ and $x = p$. Prove that $x = 0$ is stable if $p < 0$ and unstable if $p > 0$, and $x = p$ is unstable if $p < 0$ and stable if $p > 0$. Such a point $p = 0$, where there is an exchange of stability in the branches of the steady points, is called a **transcritical** bifurcation.

Problem 11.3. Consider the equation

$$\frac{dx}{dt} = px - x^3.$$

Show that $x = 0$ and $x = \pm\sqrt{p}$ (for $p > 0$) are the steady states of this equation, and determine their stability. The point $p = 0$ is called a **pitchfork** bifurcation.

Figure 11.1 illustrates the above three examples.

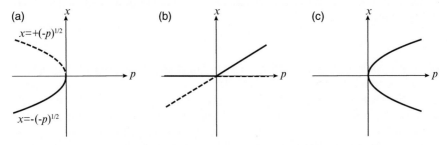

Fig. 11.1: (a) Saddle-point bifurcation diagram. (b) Transcritical bifurcation diagram. (c) Pitchfork bifurcation. Solid curves represent stable steady states, while dotted curves are unstable steady states.

Example 11.1. Consider a species x with logistic growth whose death rate is a parameter p,

$$\frac{dx}{dt} = rx\left(1 - \frac{x}{K}\right) - px. \tag{11.2}$$

It has two steady states: $x = 0$ and $x = K(1 - \frac{p}{r})$, but the latter one is biologically feasible only if $x > 0$, that is, if $p < r$. The two branches of steady points intersect at $p = r$ where exchange of stability occurs: $x = 0$ is stable if $p > r$ and unstable if $p < r$, whereas $x = K(1 - \frac{p}{r})$ is stable if $p < r$ and unstable if $p > r$. Thus transcritical bifurcation occurs at $p = r$; see Fig. 11.2.

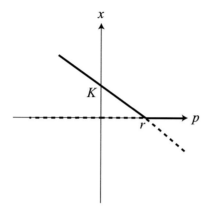

Fig. 11.2: Transcritical bifurcation diagram for Eq. (11.2). Solid lines represent stable steady states, while dotted lines are unstable steady states.

When the density of species x is very small (say $0 < x < 1$), mating becomes difficult: The probability of a male from x to meet and mate with a female from x is proportional to $x \times x$. Hence instead of growth rates

$$\frac{dx}{dt} = rx,$$

we have growth rate

$$\frac{dx}{dt} = rx^2$$

or, under constraints represented by a carrying capacity K,

$$\frac{dx}{dt} = rx^2(1 - \frac{x}{K}).$$

Example 11.2. Consider species x with dynamics

$$\frac{dx}{dt} = rx^2(1 - \frac{x}{K}) - px. \tag{11.3}$$

It has three branches of steady points given by $x = 0$ and

$$rx(1 - \frac{x}{K}) - p = 0, \text{ or } x = \frac{K}{2} \pm \sqrt{\frac{K^2}{4} - \frac{pK}{r}}.$$

In this example pitchfork bifurcation occurs at $p = \frac{r}{4}K$, as illustrated in Fig. 11.3.

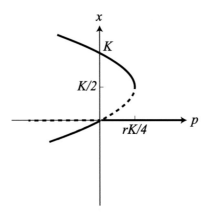

Fig. 11.3: Pitchfork bifurcation diagram for Eq. (11.3). Solid lines represent stable steady states, while dotted lines are unstable steady states.

11.1 Hopf Bifurcation

We next consider a different type of bifurcation whereby steady points bifurcate into periodic solutions; this of course must involve a dynamical system with at least two equations.

Consider the following system of two equations, with bifurcation parameter p:

$$\frac{dx_1}{dt} = px_1 - \mu x_2 - ax_1(x_1^2 + x_2^2), \qquad (11.4)$$

$$\frac{dx_2}{dt} = \mu x_1 + px_2 - ax_2(x_1^2 + x_2^2), \qquad (11.5)$$

where μ, a are positive constants. It is easily seen that the point $x_1 = x_2 = 0$ is a steady point, stable if $p < 0$ and unstable if $p > 0$. But for $p > 0$ there also exists a periodic solution,

$$x_1(t) = \sqrt{\frac{p}{a}} \cos \mu t, \quad x_2(t) = \sqrt{\frac{p}{a}} \sin \mu t,$$

which traces the circle $x_1^2 + x_2^2 = \frac{p}{a}$ as t varies.

This type of bifurcation, which gives rise to periodic solutions, is called **Hopf bifurcation**. Fig. 11.4 illustrates the periodic solutions which arise in the Hopf bifurcation. Note that the Jacobian matrix J at the $(0,0)$, where the bifurcation occurs, is given by

$$J = \begin{pmatrix} p & -\mu \\ \mu & p \end{pmatrix},$$

and the characteristic equation is

$$(p - \lambda)^2 + \mu^2 = 0,$$

so that the eigenvalues are

$$\lambda = p \pm i\mu.$$

As p crosses from $p < 0$ to $p > 0$, the two eigenvalues, at $p = 0$, become pure imaginary numbers. It is this behavior of the eigenvalues of the Jacobian matrix that gives rise to the periodic solutions. In fact, the bifurcation behavior in the example of the system (11.4)–(11.5) is a special case of the following theorem.

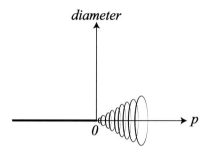

Fig. 11.4: Hopf bifurcation for Eq. (11.4)–(11.5): periodic solutions with increasing diameter $\sqrt{\frac{p}{a}}$.

Theorem 11.1. (**Hopf Bifurcation**) *Consider the system*

$$\frac{dx}{dt} = f(x, y, p), \quad \frac{dy}{dt} = g(x, y, p). \tag{11.6}$$

Assume that for all p in some interval there exists a steady state $(x^s(p), y^s(p))$, and that the two eigenvalues of the Jacobian matrix (evaluated at the steady state) are complex numbers $\lambda_1(p) = \alpha(p) + i\beta(p)$ and $\lambda_2(p) = \alpha(p) - i\beta(p)$. Assume also that

$$\alpha(p_0) = 0, \quad \beta(p_0) \neq 0 \quad and \quad \frac{d\alpha}{dp}(p_0) \neq 0.$$

Then one of the three cases must occur:

1. *there is an interval $p_0 < p < c_1$ such that for any p in this interval there exists a unique periodic orbit containing $(x^s(p_0), y^s(p_0))$ in its interior and having a diameter proportional to $|p - p_0|^{1/2}$;*
2. *there is an interval $c_2 < p < p_0$ such that for any p in this interval there exists a unique periodic orbit as in case (1);*
3. *for $p = p_0$ there exist infinitely many orbits surrounding $(x^s(p_0), y^s(p_0))$ with diameters decreasing to zero.*

A proof of Theorem 11.1 can be found, for instance, in Reference [6].

For the special system (11.4)–(11.5), $p_0 = 0$, $(x^s(p_0), y^s(p_0)) = (0, 0)$, $\alpha(p) = p$, $\beta(p) = \mu$, and case 1 occurs with $p > 0$ if $a > 0$ as shown above, case 2 occurs with $p < 0$ if $a < 0$, and case 3 occurs if $a = 0$ as illustrated in Fig. 3.1(F).

Example 11.3. We consider a model of herbivore–plant interaction. The plant P has logistic growth with carrying capacity K, and the herbivore H has eating capacity σ, which we take as the bifurcation parameter. Then

$$\frac{dP}{dt} = rP(1 - \frac{P}{K}) - \sigma\frac{P}{1+P}H,$$
$$\frac{dH}{dt} = \gamma\sigma\frac{PH}{1+P} - \mu H,$$

where γ is the yield constant and μ is the death rate of the herbivore. Rewriting these equations in the form

$$\frac{dP}{dt} = P[r(1 - \frac{P}{K}) - \frac{\sigma H}{1+P}],$$
$$\frac{dH}{dt} = H[\gamma\sigma\frac{P}{1+P} - \mu],$$

we easily compute the nonzero steady state

$$P = \frac{\mu}{\gamma\sigma - \mu}, \quad H = \frac{r}{\sigma}(1+P)(1 - \frac{P}{K}) = \frac{r\gamma}{\gamma\sigma - \mu}(1 - \frac{\mu}{K(\gamma\sigma - \mu)}),$$

and, by the factorization rule, the Jacobian is computed to be

$$J = \begin{pmatrix} P(-\frac{r}{K} + \frac{\sigma H}{(1+P)^2}) & -\sigma\frac{P}{1+P} \\ \frac{\gamma\sigma H}{(1+P)^2} & 0 \end{pmatrix}.$$

The characteristic equation is then

$$\lambda^2 - a\lambda + b = 0, \tag{11.7}$$

where $b = \det J(P,H) > 0$ and $a = \operatorname{trace} J(P,H)$ is given by

$$a = a(\sigma) = P(-\frac{r}{K} + \frac{\sigma H}{(1+P)^2}).$$

We compute

$$\sigma H = \frac{r\gamma[K(\gamma\sigma - \mu) - \mu]}{K(\gamma\sigma - \mu)^2}\sigma, \quad 1+P = \frac{\gamma\sigma}{\gamma\sigma - \mu}.$$

Hence

$$\frac{\sigma H}{(1+P)^2} = \frac{r[K(\gamma\sigma - \mu) - \mu]}{K\gamma\sigma}.$$

and

$$a(\sigma) = \frac{Pr}{K}\{-1 + \frac{1}{\gamma\sigma}[K(\gamma\sigma - \mu) - \mu]\}$$

$$= \frac{Pr}{K\gamma\sigma}[(K-1)\gamma\sigma - (K+1)\mu].$$

It follows that $a(\sigma_0) = 0$ if

$$\sigma_0 = \frac{K+1}{K-1}\frac{\mu}{\gamma}$$

and

$$\frac{da}{d\sigma}\bigg|_{\sigma=\sigma_0} = \frac{Pr}{K\gamma\sigma_0}(K-1)\gamma > 0 \quad \text{if } K > 1,$$

so that $a(\sigma) < 0$ if $\sigma < \sigma_0$.

We now observe that the points P, H are both positive if

$$\sigma > \frac{K+1}{K}\frac{\mu}{\gamma},$$

and σ_0 satisfies this inequality since

$$\frac{K+1}{K-1} > \frac{K+1}{K}.$$

We conclude that

$$a(\sigma) < 0 \quad \text{if} \quad \frac{K+1}{K}\frac{\mu}{\gamma} < \sigma < \sigma_0,$$

$$a(\sigma_0) = 0, \quad \frac{d}{d\sigma}a(\sigma_0) > 0.$$

Since the eigenvalues of (11.7) are $\lambda_{1,2} = \frac{a}{2} \pm \sqrt{\frac{a^2}{4} - b}$ and $b > 0$, we see, using Theorem 11.1, that Hopf bifurcation occurs at $\sigma = \sigma_0$. Thus as σ increases to σ_0 the stable equilibrium $(P(\sigma), H(\sigma))$ becomes unstable and, instead, the dynamics of the herbivore–plant model develops periodic solutions with diameters which increase with $|\sigma - \sigma_0|$. Thus both plant and herbivore will coexist, and their populations will vary 'seasonally.'

11.2 Neuronal Oscillations

Neuronal oscillations are periodic electrical oscillations along the axon of the neurons, and some simplified models represent them in the form

$$\frac{dv}{dt} = f(v) - w + I,$$

$$\frac{dw}{dt} = \varepsilon(\gamma v - w),$$

where I is the applied current, arriving from dendrites, which triggers the oscillations. The function $f(v)$ is a cubic polynomial and ε is a small parameter. The diameter of the periodic oscillations depends on f but is independent of the parameter I. Motivated by this model we consider here the case where f is a *quadratic* polynomial, and show that this case gives rise to Hopf bifurcation, that is, to periodic oscillations which begin with small diameter as I crosses a bifurcation parameter I_0, and then increase with I, proportionally to $(I - I_0)^{1/2}$. For simplicity we take $f(v) = v^2$.

Problem 11.4. Consider a system

$$\frac{dv}{dt} = v^2 - w + I,$$

$$\frac{dw}{dt} = 2\gamma v - w,$$

where $\gamma > \frac{1}{4}$ and $0 < I < \gamma^2$. Show that the only steady state $(\bar{v}, \bar{w})$ is given by $\bar{v} = \gamma - \sqrt{\gamma^2 - I}$, $\bar{w} = 2\gamma\bar{v}$, that it is stable if $I < \gamma - \frac{1}{4}$, and that Hopf bifurcation occurs at $I = \gamma - \frac{1}{4}$.

11.3 Endangered Species

Consider species with very sparse density v, which is endangered as a result of endemic incurable disease caused by a parasite with density w. Since the population of v is spread over a large territory, mating between a male from v and female from v is proportional to $v \times v = v^2$. Hence

$$\frac{dv}{dt} = rv^2 - \alpha vw,$$

where α is the rate by which the parasite w depletes v. On the other hand, the growth of the parasite is proportional to v, so that

$$\frac{dw}{dt} = \gamma v - \beta w,$$

where β is the death rate of w. If $r\beta - \alpha\gamma \neq 0$ then the only steady point is $(\bar{v}, \bar{w}) = (0,0)$. In order to save the endangered species v from extinction, new population of the species are introduced into the territory, at density rate I, so that

$$\frac{dv}{dt} = rv^2 - \alpha vw + I.$$

This results in steady points $(\bar{v}, \bar{w})$ where $\bar{v} > 0$, $\bar{w} > 0$, and the question arises: are these points $(\bar{v}(I), \bar{w}(I))$ stable for all I?

To address this question we take, for simplicity, $r = \alpha = \beta = 1$, and $1 < \gamma < 2$ and consider I as a bifurcation parameter. Then

$$\frac{dv}{dt} = v^2 - wv + I,$$

$$\frac{dw}{dt} = \gamma v - w.$$

The only steady point is $\bar{w} = \gamma \bar{v}$, $\bar{v} = (\frac{I}{\gamma - 1})^{1/2}$, and the Jacobian matrix about $(\bar{v}, \bar{w})$ is

$$J = \begin{pmatrix} (2 - \gamma)\bar{v} & -\bar{v} \\ \gamma & -1 \end{pmatrix}.$$

Hence $\det J = 2\bar{v}(\gamma - 1) > 0$ and

$$\text{trace } J = (2 - \gamma)(\frac{I}{\gamma - 1})^{1/2} - 1 \equiv A(I),$$

where $A(I) < 0$ if $I < I_0$, $A(I) > 0$ if $I > I_0$, with

$$I_0 = \frac{\gamma - 1}{(2 - \gamma)^2}.$$

The eigenvalues of J are

$$\lambda = \sigma \pm i\tau,$$

where $\sigma = \frac{1}{2}A(I)$, $\tau = [(\frac{1}{2}A(I))^2 - 2(\gamma - 1)\bar{v}]^{1/2}$, and $\frac{d\sigma}{dI} > 0$ at $I = I_0$. Hence $(\bar{v}, \bar{w})$ is a stable steady point if $I < I_0$, and Hopf bifurcation occurs at $I = I_0$. We conclude that as I is increased, the population $\bar{v}$, in the steady state, will increase and remain stable as long as $I < I_0$; thereafter the steady point will become unstable, and the populations of v and w will oscillate periodically.

Problem 11.5. Consider the following predator–prey model with sparse prey population, x,

$$\frac{dx}{dt} = x^2(1 - x) - xy,$$

$$\frac{dy}{dt} = 4xy - 4\alpha y,$$

where $\alpha > 0$. It has an equilibrium point $(\alpha, \alpha(1 - \alpha))$ for any $0 < \alpha < 1$. Prove that the equilibrium point is stable if $\alpha > \frac{1}{2}$ and that Hopf bifurcation occurs at $\alpha = \frac{1}{2}$.

The biological interpretation is that if the predator death rate is smaller than 2 then both predator and prey coexist in steady state, but if the predator death rate exceeds 2 then both predator and prey still coexist and their densities vary periodically.

11.4 Numerical Simulations

To plot the bifurcation diagram, one needs to scan through the parameter space and solve the ODEs for those parameters. If we would like to plot the bifurcation diagram for

$$\frac{dx}{dt} = f(x, p),$$

the first step is to plot the nullcline on the x-p plane ($f(x, p) = 0$), which corresponds to the steady states x_s under different p. Next, on the nullcline, we need to determine which part (branch) is stable and unstable. Let us consider the example

$$\frac{dx}{dt} = x^2 + p.$$

First we plot the curve of $x^2 + p = 0$ on p-x plane. In MATLAB, define the right-hand side function in a script file:

```
function y = saddlefun(x,p)
y = p + x.^2;
```

Note that p and x could be matrices in order to accommodate the matrix of discretized mesh grid on p-x space. To plot the bifurcation diagram, we create another function file called 'bifurcation.m' (see Algorithm 11.1). The input of this function is the name of the right-hand-side function of the ODE (e.g., 'saddlefun') and the ranges of x and p to plot. That is, to run this code, we should type a command similar to the following in the command window:

```
>> bifurcation('saddlefun',[-5,5],[-5,5])
```

In 'bifrurcation.m', we first discretize x-p plane with a 101×101 mesh grid (using the command 'meshgrid'). Then we try to plot the zeros of $x^2 + p$ by using the 'contour' command, as shown in Fig. 11.5(A). Next, for each p, we need to start with an initial condition x_0 which is *not* a steady state and run until we arrive near a steady state. Therefore, we avoid the x_0 too close to the nullclines, and use the rest of the points as initial conditions to do time evolution (green points in Fig. 11.5(B)). The solution will move away from the unstable branch and be attracted to the stable branch (blue circles in Fig. 11.5(C)). When we run 'bifurcation.m,' we will see Fig. 11.5(A)–(C) consecutively.

Problem 11.6. Plot the bifurcation diagram for

$$\frac{dx}{dt} = px - x^3.$$

with range $-5 \leq p \leq 5, -5 \leq x \leq 5$.

Algorithm 11.1. Plotting bifurcation diagram for $dx/dt = p + x^2$ (bifurcation.m)

```
% BIFURCATION(FCN,XRANGE,PRANGE) draws the bifurcation diagram
% for the function FCN over the specified x and p ranges.
% FCN is a handle to a user-defined function that takes as
% arguments a variable x and a parameter p.
% XRANGE is a row vector of the form [XMIN XMAX].
% PRANGE is a row vector of the form [PMIN PMAX].
% Example: % bifur(@saddlefun,[-5 5],[-5 5]);
% where saddlefun is a user-defined function of the form
%
% function y=saddlefun(x,p)
% y=p+x.2;
%
% BIFURCATION(FCN,XRANGE,PRANGE) draws the bifurcation diagram
% for the function FCN over the specified x and p ranges.
% FCN is a handle to a user-defined function that takes as
% arguments a variable x and a parameter p.
% XRANGE is a row vector of the form [XMIN XMAX].
% PRANGE is a row vector of the form [PMIN PMAX].
% Example: % bifur(@saddlefun,[-5 5],[-5 5]);
% where saddlefun is a user-defined function of the form
%
% function y=saddlefun(x,p)
% y=p+x.^2;
%

function bifurcation(fcn,xrange,prange)
nn = 100; % number of points plotted in each range
p1 = [prange(1):(prange(2)-prange(1))/nn:prange(2)]; % sample points in p
x1 = [xrange(1):(xrange(2)-xrange(1))/nn:xrange(2)]; % sample points in x

[p,x] = meshgrid(p1,x1);   % generate grid points in p and x,
                           % which are matrices
fval = feval(fcn,x,p); % evaluate the points value
figure(1);
[c,h] = contour(p,x,fval,[0,0],'r'); % plot the zero contour
pause(1)   % pause then show the next figure
xlabel('p'), ylabel('x')
x = x(:); p = p(:);  % reshape the matrices to column vectors

%% find the points whose the function values are not close to zero
%  that is, points that are not stationary points
ind = find(abs(fval)>0.05*mean(abs(fval(:))));
x = x(ind); p = p(ind);
if 1
    figure(1); hold on; plot(p,x,'go') % draw the initial points
    pause(1) % pause then show the next figure
end

%% solve the ODEs until steady states from different initial conditions
for iter = 1:1000
    % forward Euler with time step 0.05
    x = x + 0.05*feval(fcn,x,p);
end
hold on; plot(p(:),x(:),'bo')
```

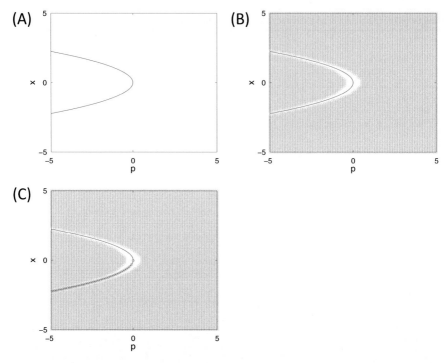

Fig. 11.5: Bifurcation diagram for saddle point bifurcation. (A) The red curve is the nullcline; (B) The green area are covered by green circles, which are non-steady-state points; (C) Solutions marked by blue are all the steady states starting with initial conditions marked by green; that is, only half of the red curve is stable.

Chapter 12
Atherosclerosis: The Risk of High Cholesterol

Arteries are blood vessels that carry oxygen-rich blood to the heart, brain, and other parts of the body. Atherosclerosis is a disease in which a plaque, a thick hard deposit of fatty material, builds up inside arteries. The plaque contains cholesterol, calcium, cells from the blood, and cells from the arterial wall. Over time the plaque grows, hardens, and narrows the artery. This reduces the flow of oxygen-rich blood, and also make it more likely to cause a blood clot, or thrombus, that will block the blood flow. A blockage formed in the coronary arteries may trigger a heart attack. A blockage formed in the carotid artery (located on each side of the neck, feeding oxygen to the brain) may cause a stroke. Atherosclerosis is the leading cause of death in the United States and worldwide, with annual deaths of 900,000 in the United States and 13 millions worldwide.

The exact cause of atherosclerosis is unknown, and in many cases there are no symptoms until an episode of heart attack or stroke occurs. There are however risk factors which contribute to the disease, namely, high cholesterol, heavy smoking, and hypertension. In this chapter we focus on the risk associated with high cholesterol, and use mathematics to quantify this risk.

Cholesterol is a protein that each cell in our body needs. But cholesterol does not dissolve in blood and must therefore be transported in the blood stream. It is transported by carrier called **lipoprotein**, made of fat (lipid) and protein. There are two types of lipoproteins that carry the cholesterol to and from cells. They are called: low-density lipoproteins, LDL, and high-density lipoproteins, HDL. The ratio of protein to cholesterol is low in the LDL and high in the HDL. The LDL are 'bad' cholesterols, and the HDL are 'good' cholesterols. The LDL contribute to plaque growth and the HDL reduce the plaque by removing the LDL from the plaque, and back to the liver, where it is broken down for recycling or for secretion from the body.

Electronic supplementary material The online version of this chapter (doi: 10.1007/978-3-319-29638-8_12) contains supplementary material, which is available to authorized users.

The level of cholesterol in the blood is measured in units of 10^{-5}g/cm^3. The American Heart Association (AHA) established guidelines regarding the atherosclerosis risk associated with the levels of LDL and HDL in the blood. For example, LDL = 190, HDL = 40 is high risk, and LDL = 110, HDL = 50 is risk free. These numbers represent concentrations in units of 10^{-5} g/cm^3.

The AHA guidelines are based on epidemiological studies that are periodically adjusted. But it would be important also to understand the mechanism of the growth of a plaque, so that we could improve the risk assessment of an artery blockage, and perhaps even develop drugs that slow or block the growth of a plaque. In this chapter, we develop a simple mathematical model which explains how the risk of plaque's growth is associated with the levels of LDL and HDL

We introduce, as parameters

$$L_0 = \text{concentration of LDL in blood,}$$
$$H_0 = \text{concentration of HDL in blood,}$$

and wish to determine, based on (L_0, H_0), whether a plaque will grow or shrink. To do that we need to understand how a plaque is formed.

The artery wall consists of three layers. The inner layer, called **intima**, is very thin and is made up of **endothelial** cells which form a barrier so that blood cannot leak out of the artery. The middle layer, called **media**, includes smooth muscle cells that enable the wall to expand (and then shrink) as blood is pumped out of the heart, 60 to 70 times per minute. The third layer, called **adventitia**, provides general support to the blood vessel. Fig. 12.1 shows an artery wall structure.

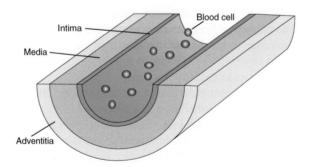

Fig. 12.1: Artery wall structure.

Free radicals are molecules or ions that have unpaired valence electrons, and are therefore highly reactive in many chemical processes in our body; they play useful role in metabolic processes. **Macrophages** are cells of the immune system that travel around the body and engulf and digest foreign particles, cellular debris, and invading microorganisms.

As a result of blood pressure, a small damage may occur in the artery wall and cholesterols from the blood may then leak out. When LDL enter the intima, they immediately become oxidized by radicals. Macrophages from the blood then move into the intima and engulf the oxidized LDL. The fat-laden macrophages saturated with oxidized LDL are called **foam cells**.

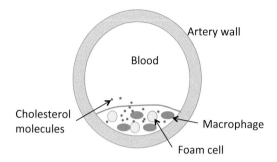

Fig. 12.2: Cross section of a plaque in an artery.

In our mathematical model we assume that the plaque consists mainly of macrophages and foam cells. This is a simplification, since also other cells are involved, for instance smooth muscle cells which move from the media into the intima. Figure 12.2 shows a cross section of a plaque in the artery.

Our model will include the following variables in units of g/cm^3:

- Macrophage density, M,
- Foam cell density, F,
- 'Bad' cholesterol concentration, LDL or L,
- 'Good' cholesterol concentration, HDL or H.

We shall not distinguish between LDL and oxidized LDL. The equation for LDL is the following:

$$\frac{dL}{dt} = L_0 - k_1 M \frac{L}{K_1 + L} - r_1 L. \tag{12.1}$$

The first term on the right-hand side, L_0, is the production rate of LDL concentration in the blood. The second term represents the ingestion of LDL by macrophages, which is described by the Michaelis-Menten formula; recall that similar 'ingestion' terms were used in the chemostat model (Eq. (8.2)) and in the plant-herbivore model (Example 11.3 in Chapter 11). The last term in Eq. (12.1) is the degradation of LDL.

In a similar way we write the equation for HDL:

$$\frac{dH}{dt} = H_0 - k_2 H \frac{F}{K_2 + F} - r_2 H. \tag{12.2}$$

Here the second term on the right-hand side is interpreted as follows: HDL is being absorbed by a foam cell (more precisely, it forms a complex with a membrane protein of a foam cell) and this initiates a process that empties out the oxidized LDL from the foam cell. The foam cells return to become a macrophage, while the emptied-out oxidized LDL are removed from the plaque and are transported back (by the blood) to the liver for recycling or secretion from the body. We note that when H forms a complex with a receptor protein on F, it takes some time for the receptor to again become free. Thus the receptor 'recycling' time limits the ability of F to react to H, and this explains why we used $k_2HF/(K_2+F)$ instead of k_2HF in Eq. (12.2): i.e., the factor $\frac{1}{K_2+F}$ accounts for the limited rate of receptor recycling.

The equations for macrophages and foam cells are

$$\frac{dM}{dt} = -k_1M\frac{L}{K_1+L} + k_2H\frac{F}{K_2+F} + \lambda\frac{ML}{H-\delta} - \mu_1M, \qquad (12.3)$$

$$\frac{dF}{dt} = k_1M\frac{L}{K_1+L} - k_2H\frac{F}{K_2+F} - \mu_2F. \qquad (12.4)$$

The first two terms or the right-hand sides of Eqs. (12.3)–(12.4), account for the exchanges between macrophages and foam cells, as already explained above. The terms μ_1M, μ_2F represent the natural deaths of macrophages and foam cells. The remaining term that needs explanation is $\lambda ML/(H-\delta)$. The oxidized LDL in the plaque triggers infiltration of macrophages from the blood into the plaque, and this is accounted by a factor $\tilde{\lambda}\tilde{M}L$, where $\tilde{\lambda}$ is a constant coefficient and $\tilde{M}$ is the concentration of macrophages outside but near the plaque; for simplicity we assume that $\tilde{M}$ is proportional to M, hence the factor λML. On the other hand, the HDL are oxidized by radicals (as are the LDL) and this reduces the amount of radicals available to oxidize LDL. In this sense H acts as an inhibitor, which restricts the effect of λML by a factor $1/(H-\delta)$, where $\delta>0$ but is small relative to H; note that H is an inhibitor only if $H>1+\delta$. We wish to solve the system of equations (12.1)–(12.4) and compute the weight of the plaque

$$w(t) = M(t) + F(t)$$

at time t; the weight of the cholesterol is negligible. We take initial values

$$L = L_0, \ H = H_0, \ F = 0, \ M = M_0 = 5\times10^{-4}\text{g/cm}^3.$$

We set

$$R(t) = \frac{w(t)}{w(0)} = \frac{w(t)}{M_0}$$

so that $R(0)=1$.

Given cholesterol level (L_0,H_0), we wish to determine whether $R(t)$ will increase, indicating risk of atherosclerosis, or decrease which means risk-free of atherosclerosis. We shall derive some results from the mathematical model by analysis and, later on, by simulations.

Auxiliary Result 1. From Eq. (12.1) we deduce the inequality

$$\frac{dL}{dt} \le L_0 - r_1L;$$

this is the same inequality as in (2.11) with $\mu = r_1, b = L_0$. Recalling Theorem 2.3 we deduce that for any small $\varepsilon > 0$,

$$L(t) < \frac{L_0}{r_1} + \varepsilon \quad \text{if } t \text{ is large,} \tag{12.5}$$

say, if $t > T_\varepsilon$.

Auxiliary Result 2. From Eq. (12.2) we deduce that

$$\frac{dH}{dt} \geq H_0 - k_2 H - r_2 H = H_0 - \frac{1}{\gamma} H,$$

where

$$\gamma = \frac{1}{r_2 + k_2}. \tag{12.6}$$

Using Theorem 2.3 we conclude that for any small $\varepsilon > 0$,

$$H(t) \geq \gamma H_0 - \varepsilon \quad \text{if } t \text{ is large,} \tag{12.7}$$

say, if $t > T_\varepsilon$ with the same T_ε as in the previous auxiliary result.

In the sequel we assume that

$$\mu_1 = \mu_2 = \mu, \quad \gamma H_0 > \delta. \tag{12.8}$$

Theorem 12.1. *If (12.8) holds and*

$$\frac{\lambda L_0 / r_1}{\gamma H_0 - \delta} < \mu, \tag{12.9}$$

then $R(t) \to 0$ as $t \to \infty$.

Before proving this assertion we note that it has been recognized in recent years that the risk of atherosclerosis depends on the ratio of L_0 to H_0, rather than on the level of each of them separately. Theorem 12.1 shows that although our model (12.1)–(12.4) is very simple, it can nevertheless demonstrate (if we drop δ) that low ratio of L_0/H_0 ensures risk-free of atherosclerosis.

Proof. (of Theorem 12.1) To prove the theorem we add equations (12.3), (12.4). Then, using (12.7), (12.8), we get

$$\frac{dw}{dt} = \frac{\lambda ML}{H - \delta} - \mu w < \frac{\lambda ML}{\gamma H_0 - \delta - \varepsilon} - \mu w$$

if $t > T_\varepsilon$, where $\gamma H_0 - \delta - \varepsilon > 0$ if ε is chosen small enough. Since $M \leq M + F = w$ and L is bounded as in (12.5), we conclude that

$$\frac{dw}{dt} \leq \left(\frac{\lambda (L_0 / r_1 + \varepsilon)}{\gamma H_0 - \delta - \varepsilon} - \mu \right) w. \tag{12.10}$$

One can verify by direct computation that

$$\frac{1}{A - \varepsilon} < \frac{1}{A} + C\varepsilon \tag{12.11}$$

for any positive constant A, if $C = \frac{1}{A} + \frac{1}{A^2}$ and ε is small, say $C\varepsilon < 1$. Hence the right-hand side of (12.10) is less than

$$[\frac{\lambda(L_0/r_1 + \varepsilon)}{\gamma H_0 - \delta}(1 + C\varepsilon) - \mu]w \leq [(\frac{\lambda L_0/r_1}{\gamma H_0 - \delta} + \frac{\lambda \varepsilon}{\gamma H_0 - \delta})(1 + C\varepsilon) - \mu]w.$$

It follows that

$$\frac{dw}{dt} \leq [\frac{\lambda L_0/r_1}{\gamma H_0 - \delta} - \mu + C_0 \varepsilon]w$$

for some constant C_0, provided $t > T_\varepsilon$. We now use the assumption (12.9) and choose ε sufficiently small to conclude that

$$\frac{dw}{dt} \leq -\alpha w$$

for some $\alpha > 0$ and $t > T_\varepsilon$. Hence

$$w(t) \leq w(T_\varepsilon)e^{-\alpha(t - T_\varepsilon)}$$

and $R(t) \to 0$ as $t \to \infty$.

Suppose the removal mechanism of LDL from foam cells is failing, which means that k_2 is very small. If also the degradation rate, r_1, of LDL is small then we expect that the plaque will not shrink. The following problem asserts that, indeed, a stationary plaque does exist but, for simplicity, we take $k_2 = 0$.

Problem 12.1. Assume that $k_2 = 0$. Prove that if $H_0 - \delta r_2 > 0$, r_1 is sufficiently small and λ is sufficiently large then there exists a unique stationary plaque $(\bar{L}, \bar{H}, \bar{M}, \bar{F})$ with $\bar{M} > 0$, $\bar{F} > 0$.

Problem 12.2. Determine whether the steady point $(\bar{L}, \bar{H}, \bar{M}, \bar{F})$ is asymptotically stable.

The model of atherosclerosis in this chapter is a simplified version of the model from the article [7].

12.1 Numerical Simulations

We wish to compute $R(t)$ for $0 < t < T$, say $T = 300$ days. We say that (L_0, H_0) is in the **risk zone** if $R(T) > 1$, and in the **risk-free zone** if $R(T) < 1$. In the following simulations we use the following parameters: $k_1 = 1.4$/day, $k_2 = 10$/day, $K_1 = 10^{-2}$g/cm^3, $K_2 = 0.5$g/cm^3, $\mu_1 = 0.003$/day, $\mu_2 = 0.005$/day, $r_1 = 2.4 \times 10^{-5}$/day, $r_2 = 5.5 \times 10^{-7}$/day, $\lambda = 2.57 \times 10^{-3}$/day, $\delta = 2.54 \times 10^{-5}$g/cm^3. The initial value for M is $M_0 = 5 \times 10^{-4}$g/cm^3.

Problem 12.3. Compute $R(300)$ (300 days) for the initial values $(L_0, H_0) =$ $(100, 60)$, $(150, 50)$, and $(200, 40)$. Note that L_0 and H_0 are in units of 10^{-5}g/cm^3. A sample code can be found in Algorithm 12.1 and 12.2.

Algorithm 12.1. Main file for simulating Problem 12.3 (main_atherosclerosis.m)

```
%%% This code is to simulate Problem 12.3
global L_0 H_0 k_1 k_2 K_1 K_2 r_1 r_2 lambda delta mu_1 mu_2

%% parameters
k_1 = 1.4;              % /day
k_2 = 10;              % /day
K_1 = 10^(-2);         % g/cm^3
K_2 = 0.5;             % g/cm^3
mu_1 = 0.003;          % /day
mu_2 = 0.005;          % /day
r_1 = 2.4*10^(-5);     % /day
r_2 = 5.5*10^(-7);     % /day
lambda = 2.57*10^(-3); % day
delta = 2.54*10^(-5);  % /day

M_0 = 5*10^(-4);       % g/cm^3
L_0 = 200*10^(-5);     % g/cm^3
H_0 = 40*10^(-5);      % g/cm^3

%% initial conditions
z_ini = [L_0, H_0, M_0, 0];
tspan = [0,300];

%% solve ODEs
[t,z] = ode15s('fun_atherosclerosis',tspan,z_ini);

w = z(:,3) + z(:,4);
R = w./M_0;

%% Plot
% plot 4 subplots for each species
figure(1)
labelvec = {'L','H','M','F'};
for i = 1 : 4
    subplot(2,2,i)
    plot(t,z(:,i))
    xlabel('t'), ylabel(labelvec(i))
end

% plot time versus R
figure(2)
plot(t,R,'g'), hold on
xlabel('t'); ylabel('R')
title(['L0 = ' num2str(L_0), '  H0 =' num2str(H_0)] )
```

Algorithm 12.2. fun_atherosclerosis.m

```
function dz = fun_atherosclerosis(t,z)
global L_0 k_1 K_1 r_1 H_0 k_2 K_2 r_2 lambda delta mu_1 mu_2

dz = zeros(4,1);

L = z(1);
H = z(2);
M = z(3);
F = z(4);

dz(1) = L_0 - k_1*M*L/(K_1+L) - r_1*L;
dz(2) = H_0 - k_2*H*F/(K_2+F) - r_2*H;
dz(3) = - k_1*M*L/(K_1+L) + k_2*H*F/(K_2+F) ...
            + lambda*M*L/(H-delta) - mu_1*M;
dz(4) = k_1*M*L/(K_1+L) - k_2*H*F/(K_2+F) - mu_2*F;
```

Chapter 13
Cancer-Immune Interaction

An abnormally new growth of tissue with cells that grow more rapidly than normal cells and has no physiological function is called a neoplasm or a tumor. The abnormally rapidly growing cells compete with normal cells for space and nutrients. When the new growth is localized, it is called a benign tumor. When a tumor in tissue has reached a size of several millimeters it requires a large supply of nutrients, for otherwise it can no longer grow. Until reaching this stage the tumor is called **avascular**. Avascular tumors that reached the stage where they require new supply of nutrients try to induce the formation of new blood vessels (**angiogenesis**) and direct their movement toward them. They do so by secreting **vascular endothelial growth factor** (VEGF) and, if successful, the tumors become **vascular**. As a tumor continues to grow some of its cells may break away and travel to other parts of the body through the bloodstream or the lymph system. **Metastatic cancer** is a tumor that spread from the original location where it started to other parts of the body. Metastatic cancer is also called **malignant cancer**, or, briefly, **cancer**, although people often use the words tumor and cancer interchangeably. Most cancer deaths are due to metastasized cancer.

Cancer is a disease of tissue growth failure, and it is the result of normal cells transforming into cancer cells because of mutations in genes that regulate cell growth and differentiation. In the context of cancer, these genes are classified either as **oncogenes** or **tumor suppressor genes**. Oncogenes are genes that promote cell growth and reproduction. Tumor suppressor genes are genes that inhibit cell division and survival. Malignant transformation occurs when oncogenes become overexpressed compared to normal oncogenes, or when tumor suppressor genes become underexpressed, or disabled. Typically a transformation of a normal cell to a tumor cell occurs after not one but several gene mutations.

Electronic supplementary material The online version of this chapter (doi: 10.1007/978-3-319-29638-8_13) contains supplementary material, which is available to authorized users.

It is commonly believed that most mutations leading to cancer are due to external conditions, such as smoking, dietary factors, environmental pollutants, exposure to radiation, and certain infections. But some mutations are hereditary.

There are more than one hundred known types of human cancer, broadly categorized according to the tissue of origin. **Carcinomas** begin with epithelial cells; **sarcomas** arise from connective tissues, muscles, and vasculature; **leukemias** and **lymphomas** are cancers of the hematopoietic (blood) and immune system, respectively; **gliomas** are cancers of the central nervous system, including the brain; **retinoblastomas** are cancers of the eyes.

The most common causes of cancer-related death in the United States are lung, colorectal, breast (for women), and prostate (for men), and pancreatic cancers. Malignancy typically induces moderate cellular immune response. But cancer cells try to evade the immune response by inducing favorable changes in phenotype of immune cells. The interaction between cancer cells and the immune system is complex, and it affects the efficacy of chemotherapeutic drugs. In order to determine the efficacy of anti-cancer drugs, we need to develop a mathematical model of cancer-immune interaction and then use it to evaluate the efficacy of various drugs; this is the aim of the present chapter.

We begin with a few facts that are needed in order to build the mathematical model. An important class of immune cells that confront a tumor are T cells. Another type of cells are macrophage, which we already met in Chapter 12. Here we distinguish between two phenotypes: pro-inflammatory macrophages M_1 and anti-inflammatory macrophages M_2. M_1 macrophages produce an inflammatory cytokine, called interleukin IL-12, and M_2 macrophages produce an anti-inflammatory cytokine, called interleukin IL-10. IL-12 activates T cells, whereas IL-10 inhibits their activation. Activated T cells kill tumor cells. In order to evade the immune system, cancer cells produce transforming growth factor β (TGF-β) that attaches to the membrane of M_1 macrophages and starts a process that changes their phenotype to M_2 macrophages, resulting in reduced killing of cancer cells by T cells. Figure 13.1 is a schematics of the cancer-immune interaction described above.

Based on Fig. 13.1 we can write down the following equations for the cells:

$$\frac{dC}{dt} = \lambda_C C(1 - \frac{C}{C_0}) - \mu_C T C, \tag{13.1}$$

$$\frac{dM_1}{dt} = k_1 - \tilde{\gamma} M_1 \frac{T_\beta}{\tilde{K}_1 + T_\beta} - \mu M_1, \tag{13.2}$$

$$\frac{dM_2}{dt} = \tilde{\gamma} M_1 \frac{T_\beta}{\tilde{K}_1 + T_\beta} - \mu M_2, \tag{13.3}$$

$$\frac{dT}{dt} = \tilde{k}_T \frac{I_{12}}{\tilde{K}_2 + I_{10}} - \mu_T T. \tag{13.4}$$

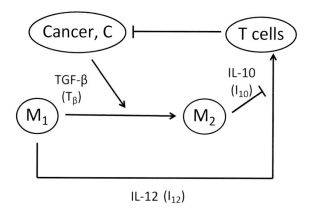

Fig. 13.1: Tumor-immune interaction. Arrow means production or activation; blocked arrow-head means inhibition or killing.

We also have the following equations for the cytokines:

$$\frac{dI_{12}}{dt} = \lambda_{12}M_1 - \mu_{12}I_{12}, \tag{13.5}$$

$$\frac{dI_{10}}{dt} = \lambda_{10}M_2 - \mu_{10}I_{10}, \tag{13.6}$$

$$\frac{dT_\beta}{dt} = \lambda_\beta C - \mu_\beta T_\beta. \tag{13.7}$$

In Eq. (13.1) we assume a logistic growth for cancer cells, and killing of cancer cells by T cells at rate μ_C. In Eq. (13.2) we assume constant production rate k_1 and death rate μ of M_1 macrophage. T_β changes the phenotype of M_1 to M_2, and this is accounted by the term $\tilde{\gamma}M_1 \frac{T_\beta}{\tilde{K}_1 + T_\beta}$, and the death rate of M_2 macrophages in Eq. (13.3) is assumed to be the same as for M_1 macrophages. In Eq. (13.4) the first term represents the activation of T cells by I_{12}, a process inhibited by I_{10} which appears in the factor $1/(\tilde{K}_2 + I_{10})$, and the second term accounts for the death of T cells at rate μ_T.

In Eqs. (13.5)–(13.7) the first term on the right-hand side is a production term by the corresponding cells, and the second term accounts for degradation. We simplify the model (13.1)–(13.7) by noting that the cytokines dynamics is much faster than the cells dynamics. Hence we may assume steady states in the equations of (13.5)–(13.7). Thus $I_{12} = constant \times M_1$, $I_{10} = constant \times M_2$ and $T_\beta = constant \times C$. Substituting these relations in Eqs. (13.2)–(13.4), the system (13.1)–(13.7) reduces to the following system of four equations:

$$\frac{dC}{dt} = \lambda_C C(1 - \frac{C}{C_0}) - \mu_C TC, \tag{13.8}$$

$$\frac{dM_1}{dt} = k_1 - \gamma M_1 \frac{C}{K_1 + C} - \mu M_1, \tag{13.9}$$

$$\frac{dM_2}{dt} = \gamma M_1 \frac{C}{K_1 + C} - \mu M_2, \tag{13.10}$$

$$\frac{dT}{dt} = k_T \frac{M_1}{K_2 + M_2} - \mu_T T, \tag{13.11}$$

with constant coefficients γ, k_T and K_1, K_2.

A common chemotherapeutic drug is TGF-β inhibitor. The effect of this drug is to increase μ_β and hence to decrease γ.

In order to determine whether TGF-β inhibitor can eradicate cancer, we shall first derive some inequalities, or estimates.

Estimate 1. For any small $\varepsilon > 0$ the following inequality holds:

$$M_1(t) > \frac{k_1}{\mu + \gamma} - \varepsilon \quad \text{for all sufficiently large } t, \tag{13.12}$$

say, $t > T_{1\varepsilon}$. Indeed, since

$$\frac{C}{K_1 + C} < 1,$$

we deduce from Eq. (13.9) that

$$\frac{dM_1}{dt} > k_1 - (\gamma + \mu)M_1.$$

The assertion (13.12) then follows by using Theorem 2.3.

Estimate 2. For any small $\varepsilon > 0$ there holds:

$$M_2(t) < \frac{\gamma k_1}{\mu^2} + \varepsilon \quad \text{for all sufficiently large } t, \tag{13.13}$$

say, $t > T_{2\varepsilon}$. Indeed, from Eq. (13.10) we get

$$\frac{dM_2}{dt} < \gamma M_1 - \mu M_2.$$

Note that from Eq. (13.9) we have

$$\frac{dM_1}{dt} < k_1 - \mu M_1,$$

which leads to

$$M_1(t) < \frac{k_1}{\mu} + \varepsilon_1.$$

Substituting this into the differential inequality for M_2, we get

$$\frac{dM_2}{dt} < \frac{\gamma k_1}{\mu} + \gamma \varepsilon_1 - \mu M_2,$$

and the inequality (13.13) then follows by using Theorem 2.3.

Estimate 3. For any small $\varepsilon > 0$ the following inequality holds:

$$T(t) > \frac{k_T}{\alpha(\gamma)} - \varepsilon \quad \text{for all sufficiently large } t, \tag{13.14}$$

say, $t > T_{3\varepsilon}$, where

$$\alpha(\gamma) = \frac{1}{k_1} \mu_T (\mu + \gamma)(K_2 + \frac{\gamma k_1}{\mu^2}). \tag{13.15}$$

To prove this result we use the inequalities (13.12), (13.13) to deduce that

$$\frac{M_1}{K_2 + M_2} > (\frac{k_1}{\mu + \gamma} - \varepsilon) \frac{1}{K_2 + \frac{\gamma k_1}{\mu^2} + \varepsilon} \tag{13.16}$$

if $t > T_{1\varepsilon} + T_{2\varepsilon}$. We next set

$$A = K_2 + \frac{\gamma k_1}{\mu^2}$$

and note, similarly to the inequality (12.11), that

$$\frac{1}{A + \varepsilon} > \frac{1}{A} - C_1 \varepsilon$$

if

$$C_1 = \frac{1}{A} + \frac{1}{A^2}$$

and ε is sufficiently small so that $C_1 \varepsilon < 1$. Using this in (13.16), we get

$$\frac{M_1}{K_2 + M_2} > (\frac{k_1}{\mu + \gamma} - \varepsilon)(\frac{1}{A} - C_1 \varepsilon)$$

$$> \frac{k_1}{\mu + \gamma} \frac{1}{A} - C_2 \varepsilon$$

for some constant C_2. If we substitute this into Eq. (13.11), we obtain the inequality

$$\frac{dT}{dt} > [k_T \frac{k_1}{\mu + \gamma} \frac{1}{A} - k_T C_2 \varepsilon] - \mu_T T,$$

for $t > T_{1\varepsilon} + T_{2\varepsilon}$, where $\frac{1}{\mu_T} \frac{k_1}{\mu + \gamma} \frac{1}{A} = \frac{1}{\alpha(\gamma)}$, by the definition of $\alpha(\gamma)$ in (13.15). We now apply again Theorem 2.3 to deduce that

$$T(t) > \frac{k_T}{\alpha(\gamma)} - \frac{1}{\mu_T} k_T C_2 \varepsilon - \varepsilon \tag{13.17}$$

if $t - (T_{1\varepsilon} + T_{2\varepsilon})$ is sufficiently large, say if $t > T_{3\varepsilon}$. We next observe that since ε can be taken to be any small positive number, we could have taken it such that also

$$\frac{1}{\mu_T} k_T C_2 \varepsilon + \varepsilon$$

is an arbitrarily small number. Hence the inequality (13.14) holds with arbitrarily small ε provided t is sufficiently large.

Recall that the effect of the anti-cancer drug TGF-β inhibitor is to decrease the parameter γ which appears in Eq. (13.10) and in (13.15). The maximum efficacy of the drug (ignoring negative side-effects) is when $\gamma = 0$, and

$$\alpha(0) = \frac{1}{k_1} \mu_T \mu K_2. \tag{13.18}$$

The following theorem gives a sufficient condition under which the treatment with TGF-β inhibitor will cure cancer.

Theorem 13.1. *If*

$$k_T \mu_C > \lambda_C \alpha(\gamma) \tag{13.19}$$

then $C(t) \to 0$ as $t \to \infty$.

Proof. Substituting (13.14) into Eq. (13.8), we get

$$\frac{dC}{dt} \le \lambda_C C - \mu_C T C < (\lambda_C - \mu_C \frac{k_T}{\alpha(\gamma)} - \varepsilon) C$$

if $t > T_{3\varepsilon}$ and, by (13.19), the right-hand side is smaller than $-\beta C$ for some $\beta > 0$, provided ε is chosen sufficiently small. It follows that

$$C(t) \le C(T_{3\varepsilon}) e^{-\beta(t - T_{3\varepsilon})} \quad \text{if } t > T_{3\varepsilon},$$

so that $C(t) \to 0$ as $t \to \infty$.

The parameter k_T in (13.9) depends on the strength of the immune system: As seen from Eq. (13.11) this parameter depends on how fast the T cells can be mobilized by macrophages to kill cancer cells. The parameter μ_C depends on how effective the T cells are in recognizing and in killing cancer cells. Thus, altogether, the product $k_T \mu_C$ represents the total strength of the immune system in fighting cancer cells. The function $\alpha(\gamma)$ depends on the efficacy of the TGF-β inhibitor, with the maximum efficacy being $\alpha(0)$. Theorem 13.1 says that drug alone may not guarantee cancer eradication: The immune system needs to be strong enough so that

$$k_T \mu_C > \frac{\lambda_C}{k_1} \mu_T \mu K_2; \tag{13.20}$$

only then (13.19) could be satisfied under treatment with γ small enough to ensure that the cancer is eradicated. We note however that the inequality (13.19) provides a rather crude sufficient condition for cancer eradication; numerically one can derive more refined sufficient condition.

If the inequality (13.20) is reversed then Theorem 13.1 cannot be applied, that is, the cancer $C(t)$ may not disappear as $t \to \infty$ even if $\gamma = 0$ (i.e., under any drug treatment with TGF-β inhibitor). Indeed, the following problem asserts that there exist steady state solutions with $C > 0$ whenever

$$k_T \mu_C < \frac{\lambda_C}{k_1} \mu_T \mu K_2. \tag{13.21}$$

Problem 13.1. Assume that (13.21) holds and that $\gamma = 0$ so that $M_2 = 0$, Eq. (13.10) drops out, and Eq. (13.9) becomes

$$\frac{dM_1}{dt} = k_1 - \mu M_1.$$

Prove that the reduced system (13.8)–(13.11) has a unique steady state $(\bar{C}, \bar{M}_1, \bar{T})$ with $\bar{C} > 0$, and $\bar{C} \to 0$ if

$$\mu_C \to \frac{\lambda_C}{k_T k_1} \mu_T \mu K_2.$$

Problem 13.2. Show that the steady state $(\bar{C}, \bar{M}_1, \bar{T})$ is stable.

If (13.21) holds and γ is not equal to zero, but it is sufficiently small, then there is a unique steady state solution of the form

$$(\bar{C} + O(\gamma), \bar{M}_1 + O(\gamma), M_2 = O(\gamma), \bar{T} + O(\gamma))$$

where the functions $O(\gamma)$ depends on γ and satisfy: $|O(\gamma)| \leq const \times \gamma$. Furthermore, writing explicitly the Jacobian matrix at this steady point, one can check that all the eigenvalues of the characteristic polynomial are negative, thus ensuring the stability of the steady point.

The model on cancer-immune interaction in this chapter is a simplified version of the model from article [8].

13.1 Numerical Simulations

We would like to use the model (13.8)–(13.11) to investigate how effective the anti-cancer drugs are. In the following problems the parameters are given as follows: $\lambda_C = 10^{-2}$/day, $\mu_C = 10^{-5}$/cell/day, $C_0 = 10^6$ cell/cm^3, $\mu = 0.3$/day, $k_1 = 3000$ cell/cm^3/day, $\gamma = 200$/day, $k_T = 3300$/cell/day, $K_1 = 0.05C_0$, $K_2 = 10^5$ cell/cm^3, $\mu_T = 0.2$/day.

Problem 13.3. Solve the model (13.8)–(13.11) under the initial conditions $C(0) = 10^2$cell/cm^3, $M_1(0) = 5 \times 10^4$cells/cm^3, $M_2(0) = 0$, $T(0) = 0$, for $0 \le t \le 60$ days. Sample codes are shown in Algorithms 13.1 and 13.2.

Problem 13.4. Repeat the calculation with γ replaces by γ/A, $A = 2, 5, 10, 100$ and draw the four profiles of $C(t)$, $0 \le t \le 60$. These profiles illustrate how effective the drug TGF-β inhibitor is in slowing down the growth of cancer.

The killing of cancer cells in T cells is limited by a protein PD-1 which 'restrains' the toxic activity of the T cells. A recently approved drug (anti-PDL1) blocks the activity of PD-1 and thus increases μ_C.

Problem 13.5. Repeat Problem 13.4 with

1. μ_C replaced by $\frac{\mu_C}{10}$.
2. μ_C replaced by $10\mu_C$.

Algorithm 13.1. Main file for Problem 13.3 (main_cancer_immune.m)

```
% This code is to simulate Problem 13.3
% It will generate 4 curves corresponding to different gammas

clear all
close all

global lambda_c C_0 mu_c k_1 gamma K_1 K_2 k_T mu mu_T

%% parameters
lambda_c = 10^(-2); % /day
mu_c     = 10^(-5); % /cell/day
C_0      = 10^6;    % cell/cm^3
mu       = 0.3;     % /day
k_1      = 3000;    % cell/cm^3/day
gamma    = 200;     % /day
mu_T     = 0.2;     % /day
k_T      = 3300;      % /cell/day
K_1      = 0.05*C_0; % cell/cm^3
K_2      = 10^5;      % cell/cm^3

%% initial conditions
C   = 10^2;          % cell/cm^3
M_1 = 5*10^4;        % cell/cm^3
M_2 = 0;
T   = 0;

z_ini = [C M_1 M_2 T];  % initial conditions
tspan = [0,60];

%% ODE solver
[t,z] = ode15s('fun_cancer_immune',tspan,z_ini);

%% plot
tvec = {'C','M_1','M_2','T'};  % array of strings for ylabels
for i = 1 : 4
    subplot(2,2,i)
    plot(t,z(:,i)), hold on
    xlabel('t'),ylabel(tvec(i))
end
```

Algorithm 13.2. fun_cancer_immune.m

```
function dz = fun_cancer_immune(t,z)
global lambda_c C_0 mu_c k_1 gamma K_1 K_2 k_T mu mu_T

dz = zeros(4,1);

C    = z(1);
M_1  = z(2);
M_2  = z(3);
T    = z(4);

dz(1)  = lambda_c*C*(1-C/C_0) - mu_c*T*C;
dz(2)  = k_1 - gamma*M_1*C/(K_1+C) - mu*M_1;
dz(3)  = gamma*M_1*C/(K_1+C) - mu*M_2;
dz(4)  = k_T*M_1/(K_2+M_2) - mu_T*T;
```

Chapter 14
Cancer Therapy

There are many drugs that are used in the treatment of cancer; some drugs kill cancer cells directly while others change the cancer microenvironment to make it resistant to cancer cells growth. In Chapter 13, we considered a drug, TGF-β inhibitor, which changes the macrophage phenotype, thereby enabling the immune system to kill cancer cells more effectively.

In this chapter we consider two entirely different kinds of anti-cancer drugs. The first one blocks the activity of vascular endothelial growth factor (VEGF), and the second one uses virus to kill cancer cells.

14.0.1 VEGF Receptor Inhibitor

In order to continue to grow abnormally, the tumor requires increasing amounts of oxygen (and other nutrients) from the blood. So the tumor secrets VEGF which attracts endothelial cells that form the inner lining of the blood vessels' wall, thereby leading to the formation of new blood vessels (**angiogenesis**) which deliver oxygen (and other nutrients) to the tumor. To model this process we introduce the following variables:

$$c = \text{density of tumor cells,}$$
$$e = \text{density of endothelial cells,}$$
$$h = \text{concentration of VEGF,}$$
$$w = \text{concentration of oxygen,}$$

We assume logistic growth

$$\tilde{\lambda}_1 c \left(1 - \frac{c}{K}\right)$$

Electronic supplementary material The online version of this chapter (doi: 10.1007/978-3-319-29638-8_14) contains supplementary material, which is available to authorized users.

C.-S. Chou, A. Friedman, *Introduction to Mathematical Biology*,
Springer Undergraduate Texts in Mathematics and Technology,
DOI 10.1007/978-3-319-29638-8_14

of the tumor with $0 < c(0) < K$ initially, where K is the **carrying capacity** and $\tilde{\lambda}_1$ is the growth rate. We assume that $\tilde{\lambda}_1$ is proportional to w, $\tilde{\lambda} = constant \times w$, so that $\tilde{\lambda}_1 = 0$ if $w = 0$; if there is no oxygen then there is no growth. We also assume that w depends linearly on the density of blood vessels, which is proportional to the density of endothelial cells, so that

$$w = Be \quad \text{(B a positive constant)}. \tag{14.1}$$

Hence,

$$\frac{dc}{dt} = \lambda_1 ec(1 - \frac{c}{K}) - \mu_1 c, \quad 0 < c(0) < K. \tag{14.2}$$

Here λ_1 is a positive constant and μ_1 is the death rate of cancer cells.

Next we model the equation for VEGF by

$$\frac{dh}{dt} = \lambda_2 c - \mu_2 h, \tag{14.3}$$

where λ_2 is the production rate of VEGF by tumor cells, and μ_2 is the degradation rate.

Oxygen is decreased by consumption by cancer cells (at rate μ_3) as well as by dissipation in the tissue (at rate μ_4) so that

$$\frac{dw}{dt} = -\bar{\mu}_3 cw - \bar{\mu}_4 w.$$

and, by the identification of w with B_e in Eq. (14.1), we get

$$\frac{de}{dt} = -\mu_3 ce - \mu_4 e.$$

Assuming that endothelial cells proliferation is proportional to h, the complete equation for e then takes the form

$$\frac{de}{dt} = \lambda_3 h - \mu_3 ce - \mu_4 e, \tag{14.4}$$

where all the parameters are positive constants.

Avastin is a drug that inhibits VEGF receptor (VEGFR) and thus blocks the activity of VEGF. We can model the effect of Avastin by replacing λ_2 in Eq. (14.3) by $\lambda_2/(1+A)$ where A is proportional to the amount of the delivered drug. Then Eq. (14.3) becomes

$$\frac{dh}{dt} = \frac{\lambda_2 c}{1+A} - \mu_2 h. \tag{14.5}$$

We wish to explore to what extent Avastin can slow cancer growth or even eliminate cancer. To do that we first derive several estimates on $h(t)$ and $e(t)$.

Estimate 1. For any $\varepsilon > 0$ there holds:

$$h(t) < \frac{\lambda_2 K}{\mu_2(1+A)} + \varepsilon \quad \text{if } t \text{ is sufficiently large}, \tag{14.6}$$

say, $t > T_{1\varepsilon}$. To prove it we first note that

$$c(t) < K \quad \text{for all } t > 0. \tag{14.7}$$

Indeed, since $c(t) < K$ if $t = 0$, if the assertion (14.7) is not true then there is a smallest time $t = t_0$ such that $c(t) < K$ if $t < t_0$ and $c(t_0) = K$. Hence $dc(t_0)/dt \geq 0$. However, from Eq. (14.2) we get

$$\frac{dc}{dt}(t_0) = -\mu_1 c(t_0) < 0$$

which is a contradiction.

Substituting (14.7) into Eq. (14.5) we get

$$\frac{dh}{dt} \leq \frac{\lambda_2 K}{1+A} - \mu_2 h.$$

We can then apply Theorem 2.3 to derive the assertion (14.6).

Estimate 2. For any $\varepsilon > 0$ there holds:

$$e(t) \leq \frac{\lambda_2 \lambda_3 K}{\mu_2 \mu_4 (1+A)} + \varepsilon \quad \text{if } t \text{ is sufficiently large,} \tag{14.8}$$

say, $t > T_{1\varepsilon}$. To prove it we first note that by Eq. (14.4),

$$\frac{de}{dt} \leq \lambda_3 h - \mu_4 e.$$

Substituting (14.6) into this inequality, we get

$$\frac{de}{dt} \leq \lambda_3 \left(\frac{\lambda_2 K}{\mu_2 (1+A)} + \varepsilon \right) - \mu_4 e \quad \text{if } t > T_{1\varepsilon}.$$

Then, by Theorem 2.3,

$$e(t) < \frac{\lambda_2 \lambda_3 K}{\mu_2 \mu_4 (1+A)} + \left(\frac{\lambda_3 \varepsilon}{\mu_4} + \varepsilon \right) \tag{14.9}$$

if $t - T_{1\varepsilon}$ is sufficiently large, say $t - T_{1\varepsilon} > T_{2\varepsilon}$. The inequality (14.9) holds for any small $\varepsilon > 0$, and hence for any small $\lambda_3 \varepsilon / \mu_4 + \varepsilon$. By viewing $\lambda_3 \varepsilon / \mu_4 + \varepsilon$ as another small new 'epsilon,' the assertion (14.6) follows.

Theorem 14.1. *If A is sufficiently large so that*

$$1 + A > \frac{\lambda_1 \lambda_2 \lambda_3 K}{\mu_1 \mu_2 \mu_4} \tag{14.10}$$

then $c(t) \to 0$ as $t \to \infty$.

Proof. From Eq. (14.2), we obtain the inequality

$$\frac{dc}{dt} < \lambda_1 ec - \mu_1 c = (\lambda_1 e - \mu_1)c$$

and, by (14.8),

$$\lambda_1 e - \mu_1 < \frac{\lambda_1 \lambda_2 \lambda_3 K}{\mu_2 \mu_4 (1+A)} + \lambda_1 \varepsilon - \mu_1. \qquad (14.11)$$

The condition (14.10) implies that if ε is small enough then the right-hand side of (14.11) is smaller than a negative number $-\alpha$, provided t is large enough. Hence

$$\frac{dc}{dt} \le -\alpha c \quad \text{if } t \text{ is sufficiently large,}$$

and then $c(t) \to 0$ as $t \to \infty$.

From Theorem 14.1 we conclude that if Avastin can be administered in very large amount then the tumor will shrink to zero. However, Avastin has negative side-effects including damage to the liver, and thus can only be administered in limited amounts. In mouse experiments Avastin has been shown to cure cancer all by itself. However, in humans Avastin is typically used in combination with other chemotherapeutic drugs that are cancer specific.

From the model (14.2)–(14.4), it is reasonable to expect that if $\lambda_1, \lambda_2, \lambda_3$ are increased then the tumor will increase. In the next problem we provide an example where the tumor is in a steady state (benign tumor).

Problem 14.1. If

$$\lambda_1 \lambda_2 \lambda_3 > \mu_1 \mu_2 \mu_3,$$

then there exist two steady states (c_1, h_1, e_1), (c_2, h_2, e_2) of the system (14.2)–(14.4) provided K is sufficiently large.

A mathematical model of a disease often focuses on one major aspect of the disease. Hence we may find completely different models describing the same disease. But even when focusing on the same aspect of a disease, for example on the angiogenesis factor in cancer growth, one may develop different approaches to modeling by representing the same phenomenon in different ways. We illustrate this here by introducing another model to evaluate the effect of Avastin on cancer growth.

We denote the cancer cells density by x and model its growth as in Eq. (14.2), but with $e = 1, K = y$:

$$\frac{dx}{dt} = \lambda_1 x \left(1 - \frac{x}{y}\right) - \mu_1 x \quad (\lambda_1 > \mu_1), \qquad (14.12)$$

where the carrying capacity is taken to be the density y of the blood vessels which provide oxygen to the tumor. We model the concentration y by the equation

$$\frac{dy}{dt} = B - 2\mu y + \delta xy. \qquad (14.13)$$

Hence $B - 2\mu y$ represents that natural growth and degradation of capillaries, and δxy represents the formation of new capillaries from existing capillaries, induced by growth factors secreted by the tumor cells.

Rewriting (14.12) in the form

$$\frac{dx}{dt} = x[\lambda_1(1 - \frac{x}{y}) - \mu_1]$$

we find the following equilibrium points $(\bar{x}, \bar{y})$: $(0, \frac{B}{2\mu})$ and $\bar{x} = (1 - \frac{\mu_1}{\lambda_1})\bar{y}$, where

$$B - 2\mu\bar{y} + \delta_1\bar{y}^2 = 0, \quad \delta_1 = \delta(1 - \frac{\mu_1}{\lambda_1}).$$

Hence the two nonzero equilibrium points are

$$Z_\pm = ((1 - \frac{\mu_1}{\lambda_1})\bar{y}_\pm, \bar{y}_\pm), \tag{14.14}$$

where

$$\bar{y}_\pm = \frac{1}{\delta_1}(\mu \pm \sqrt{\mu^2 - \delta_1 B}). \tag{14.15}$$

The Jacobian matrix at $(0, \frac{B}{2\mu})$ is

$$\begin{pmatrix} \lambda_1 - \mu_1 & 0 \\ \frac{\delta B}{2\mu} & -2\mu \end{pmatrix}.$$

Since $\lambda_1 - \mu_1 > 0$, the equilibrium point $(0, \frac{B}{2\mu})$ is unstable. On the other hand the steady points $Z_\pm$ defined by (14.14) and (14.15) are biologically relevant only if $\delta_1 B < \mu^2$; if $\delta_1 B > \mu^2$ then the oxygen supply is sufficiently large and we expect the tumor to grow, rather than stabilize at an equilibrium and remain benign.

We proceed to consider the case $\delta_1 B < \mu^2$. By the factorization rule we find that

$$J(Z_\pm) = \begin{pmatrix} -\lambda_1 \frac{\bar{x}}{\bar{y}} & \lambda_1 \frac{\bar{x}^2}{\bar{y}^2} \\ \delta\bar{y} & \delta\bar{x} - 2\mu \end{pmatrix}_{Z_\pm}.$$

For the steady state to be stable we need trace $J < 0$ and $\det J > 0$, that is

$$\lambda_1 \frac{\bar{x}}{\bar{y}} + 2\mu > \delta\bar{x},$$

and

$$\lambda_1 \frac{\bar{x}}{\bar{y}}(2\mu - \delta\bar{x}) > \delta\bar{y}\frac{\lambda_1\bar{x}^2}{\bar{y}^2}.$$

These two inequalities are satisfied if and only if $\mu > \delta\bar{x}$ where $\bar{x} = (1 - \frac{\mu_1}{\lambda_1})\bar{y}_\pm$, that is,

$$\mu > \delta(1 - \frac{\mu_1}{\lambda_1})\bar{y}_\pm = \delta_1\bar{y}_\pm = \mu \pm \sqrt{\mu^2 - \delta_1 B},$$

by (14.15). Hence Z_- is a stable steady point while Z_+ is unstable.

We recall that tumors secrete VEGF which increases angiogenesis. A drug that blocks VEGF produced by the tumor, such as VEGFR-1 (e.g., Avastin), reduces δx. If δ is such that $\delta_1 B < \mu^2$ then the tumor will not grow indefinitely, and it is expected to stabilize at Z_-, where the total tumor load will be

$$\bar{x} = \frac{1}{\delta}(\mu - \sqrt{\mu^2 - \delta_1 B}).$$

Note that $\delta\bar{x}$ is decreasing with δ.

So far we modeled the tumor evolution using a logistic growth. But there are other models of tumor growth, one of the most notable introduced by **Gumpertz**. This model includes cancer cells x and growth factor γ which acts like VEGF in providing nutrients to the cancer. The system of equations for x and γ is as follows:

$$\frac{dx}{dt} = \gamma x, \tag{14.16}$$

$$\frac{d\gamma}{dt} = -\alpha\gamma, \tag{14.17}$$

where α is the depletion rate of γ.

If we substitute

$$\gamma = -\frac{1}{\alpha}\frac{d\gamma}{dt}$$

from Eq. (14.17) into Eq. (14.16), we get

$$\frac{1}{x}\frac{dx}{dt} + \frac{1}{\alpha}\frac{d\gamma}{dt} = 0$$

and, by integration,

$$\ln x + \frac{1}{\alpha}\gamma = constant = K_0, \tag{14.18}$$

where $K_0 = \ln x(0) + \frac{1}{\alpha}\gamma(0)$. Substituting γ from Eq. (14.18) into Eq. (14.16) we get the **Gumpertz equation**

$$\frac{dx}{dt} = -\alpha x \ln \frac{x}{K}. \tag{14.19}$$

We view K as the carrying capacity; it depends on the amount of nutrients available to the tumor. Hence K is a function of the concentration of blood capillaries which, as before, we denote by y. For simplicity we take $K = y$, so that

$$\frac{dx}{dt} = -\alpha x \ln \frac{x}{y}$$

and, as before, y satisfies Eq. (14.13).

Problem 14.2. The system (14.19), (14.13) has steady states

$$x = y = Z_\pm, \quad \text{where } Z_\pm = \frac{1}{\delta}(\mu \pm \sqrt{\mu^2 - \delta B}),$$

provided $\delta A < \mu^2$. Prove that the steady state $x = y = Z_-$ is stable, and that the steady state $x = y = Z_+$ is unstable.

The biological interpretation of this result is the same as for the model (14.12)–(14.13).

14.0.2 Virotherapy

We next consider anti-cancer drug which employs virus particles to kill cancer cells; such a treatment is called **virotherapy**. The virus particles are genetically modified so that they can infect cancer cells but not normal healthy cells. Such viruses are called **oncolytic viruses**. The viruses are injected directly into the tumor.

After entering a cancer cell, a virus begins to quickly replicate, and when the cancer cell dies, a large number of virus particles burst out and proceed to infect other cancer cells.

To model this process we introduce the following variables:

$x =$ number density of cancer cells,

$y =$ number density of infected cancer cells,

$n =$ number density of dead cells,

$v =$ number density of virus particles which are not contained in cancer cells.

Virotherapy is modeled by the following system of equations:

$$\begin{aligned}
\frac{dx}{dt} &= \alpha x - \beta xv, \\
\frac{dy}{dt} &= \beta xv - \delta y, \\
\frac{dn}{dt} &= \delta y - \mu n, \\
\frac{dv}{dt} &= b\delta y - \gamma v,
\end{aligned} \tag{14.20}$$

where

$\alpha =$ proliferation rate of cancer cells,

$\beta =$ rate of infection of cancer cells by viruses,

$\delta =$ death rate of infected cancer cells,

$\mu =$ removal rate of debris of dead cells,

and, finally, b is the replication number of a virus at the time of death of the infected cancer cell. Adding the first three equations of Eqs. (14.20), we get

$$\frac{d}{dt}(x+y+n) = \alpha x - \mu n. \tag{14.21}$$

We assume that the tumor is spherical with radius $R(t)$ and volume $V(t)$. Then, by (14.20) the total density of the cells, at each point of the sphere, increases at rate $\alpha x - \mu n$. The total mass of the tumor then increases at rate $(\alpha \tilde{x} - \mu \tilde{n})V(t)$, where $\tilde{x}, \tilde{n}$ are averages of x and n, respectively.

We next assume that this increase in total mass causes the tumor volume to grow proportionally, that is, by $\theta_0(dV/dt)$, for some constant θ_0. Then,

$$\theta_0 \frac{dV(t)}{dt} = (\alpha \tilde{x} - \mu \tilde{n})V(t). \tag{14.22}$$

From $V(t) = 4\pi R(t)^3/3$, we get

$$\frac{1}{V(t)}\frac{dV(t)}{dt} = \frac{3}{R(t)}\frac{dR}{dt},$$

so that, by (14.22),

$$\theta_0 \frac{3}{R}\frac{dR}{dt} = \alpha \tilde{x} - \mu \tilde{n} = \alpha \tilde{x} - \mu(\theta_0 - \tilde{x} - \tilde{y}).$$

if we assume that $\tilde{x} + \tilde{y} + \tilde{n} = \theta_0$.

Assuming also that $\tilde{x}, \tilde{y}$ satisfy the same equation as x, y, we get

$$\theta_0 \frac{3}{R}\frac{dR}{dt} = \alpha x - \mu n. \tag{14.23}$$

In experiments, viral therapy as described above was not initially successful because it failed to address the effect of the immune system. Immune cells recognize the infected cancer cells and destroy them before the virus particles get a chance to replicate to their full potential. To make virotherapy more effective the immune system must therefore be suppressed. In Problem 14.5 we extend the model (14.20)–(14.23) to include the density of the immune cells, z, and the chemotherapy P which suppresses the immune system.

Problem 14.3. Show that the system (14.20) has a steady point $(\bar{x}, \bar{y}, \bar{n}, \bar{v})$ with $\bar{x} > 0$, and determine whether it is asymptotically stable.

The mathematical model on cancer virotherapy in this chapter is a simplification of the model from article [9].

14.1 Numerical Simulations

To simulate the model (14.20)–(14.23), we provide the sample codes in Algorithms 14.1 and 14.2.

Problem 14.4. Take $\alpha = 2 \times 10^{-1}$/h, $\delta = (1/18)$/h, $\mu = (1/48)$/h, $\theta_0 = 10^6$ cells/mm^3, $\beta = 7 \times 10^{-8}$mm^3/h/virus, $\gamma = 2.5 \times 10^{-2}$/h. Compute $R(t)$ for $0 \leq t \leq 20$h, with initial conditions $x_0 = 8 \times 10^5$ cells/mm^3, $x_0 + y_0 + n_0 = \theta_0$, $y_0 = 10^5$ cells/mm^3, $v_0 = 10^6$ virus/mm^3, $R(0) = 2$ mm when $b = 50, 100, 200, 500$.

Problem 14.5. Consider the system

$$\frac{dx}{dt} = \alpha x - \beta xv,$$

$$\frac{dy}{dt} = \beta xv - kyz - \delta y,$$

$$\frac{dz}{dt} = syz - \omega z^2 - P(t)z,$$

$$\frac{dn}{dt} = kyz + \delta y - \mu n,$$

$$\frac{dv}{dt} = b\delta y - k_0 vz - \gamma v,$$

where z = number density of immune cells, $P(t)$ = immune suppressor drug, $x + y + z + n = \theta_0$, k = rate of immune cell killing infected cell, k_0 = take-up rate of virus by immune cells, s = stimulation rate of immune cells by infected cells, ω = clearing rate of immune cells. We take $P(t) = 0.5$/h, $k = 2 \times 10^{-8}$ mm^3/h/immune cell, $k_0 = 10^{-8}$ mm^3/h/immune cell, $s = 5.6 \times 10^{-7}$ mm^3/h/infected cell, $\omega = 2 \times 10^{-12}$ mm^3/h/immune cell, and all other parameters as in Problem 14.4, $z_0 = 6 \times 10^4$ cells/mm^3, and all other initial conditions as in Problem 14.4. (i) Compute $R(t)$ for $0 \leq t \leq 20$h, when $b = 50, 100, 200, 500$ and compare the results with those of Problem 14.4. (ii) Do the same when the chemotherapy dose is increased to $P(t) = 1$/h.

Algorithm 14.1. Main file for Problem 14.4 (main_cancer.m)

```
%%% This code simulates model (14.20)-(14.23).

%% define global parameters
global alpha delta mu theta_0 beta gamma b

%% starting and final time
t0 = 0; tfinal = 20;

%% paramters
alpha    = 2*10^-1;
delta    = 1/18;
mu       = 1/48;
theta_0  = 10^6;
beta     = 7*10^-8;
gamma    = 2.5* 10^-2;
b        = 50;

%% initial conditions
x0 = 8*10^5;
y0 = 10^5;
n0 = theta_0 - x0 - y0;
v0 = 10^6;
R0 = 2;
w_ini = [x0, y0, n0, v0, R0];
[t,w] = ode45('fun_cancer',[t0,tfinal],w_ini);
lablevec = ['x','y','n','v','R'];
for i = 1:5,
    subplot(2,3,i);
    plot(t,w(:,i)); hold on
    xlabel('time'), ylabel(lablevec(i))
end
```

Algorithm 14.2. fun_cancer.m

```
%%% This function is called by main_cancer.m
function dy = fun_cancer(t,w)
global alpha delta mu theta_0 beta gamma b

x = w(1);   y = w(2);   n = w(3);   v = w(4);   R = w(5);

dy(1) = alpha*x - beta*x*v;
dy(2) = beta*x*v - delta*y;
dy(3) = delta*y - mu*n;
dy(4) = b*delta*y - gamma*v;
dy(5) = R/(3*theta_0)*((alpha+mu)*x + mu*y - mu*theta_0);

dy = [dy(1);dy(2);dy(3);dy(4);dy(5)];
```

Chapter 15
Tuberculosis

Tuberculosis (TB) is an infective disease caused by Mycobacterium tuberculosis (Mtb). The bacteria is spread through the air when people who have active TB infection cough or sneeze. The bacteria attack the lungs, primarily, but can also spread and attack other parts of the body. The most common symptom of active TB infection is chronic cough with blood-tinged sputum. It is estimated that one-third of the world's population are infected with Mtb, although only 13 million chronic cases are active, and 1.5 million associated deaths occur. Treatment of TB uses antibiotics to kill the bacteria, but the treatment is not entirely effective. Vaccination in children decreases significantly the risk of infection.

TB infection in the lungs begins when inhaled Mycobacteria tuberculosis reach the pulmonary alveoli and invade into, or are ingested by, alveoli macrophages; alveoli are tiny air sacs within the lungs where exchange of oxygen and carbon dioxide takes place. It is important to determine whether infection by inhaled Mtb will develop into chronic TB. This cannot be done directly by measurements, so we shall use mathematics to address this question. In what follows we develop a mathematical model and use it to estimate the threshold of an initial infection that will develop into active TB.

We introduce the following variables:

M = number of alveolar macrophages in cm^3;

M_i = number of infected alveolar macrophages in cm^3;

B_e = number of extracellular bacteria (residing in tissue, outside macrophages) in cm^3;

B_i = number of intracellular bacteria (residing inside macrophages) in cm^3.

M satisfies the differential equation

Electronic supplementary material The online version of this chapter (doi: 10.1007/ 978-3-319-29638-8_15) contains supplementary material, which is available to authorized users.

C.-S. Chou, A. Friedman, *Introduction to Mathematical Biology*,
Springer Undergraduate Texts in Mathematics and Technology,
DOI 10.1007/978-3-319-29638-8_15

$$\frac{dM}{dt} = \mu_M - \lambda_1 M \frac{B_e}{K+B_e} - d_M M. \tag{15.1}$$

Here μ_M is the production rate of M and d_M is the death rate when there is no infection; in steady state, $\mu_M = d_M M_0$ where M_0 is the number of macrophages in cm^3 in healthy lungs. The second term on the right-hand side of Eq. (15.1) represents the ingestion of bacteria by macrophages, modeled by the Michaelis-Menten formula, which turns M into M_i.

The infected macrophages satisfy the equation

$$\frac{dM_i}{dt} = \lambda_1 M \frac{B_e}{K+B_e} - \lambda_2 M_i \frac{B_i^2}{B_i^2 + (NM_i)^2} - d_{M_i} M_i. \tag{15.2}$$

The first term on the right-hand side comes from macrophages ingesting extracellular bacteria, and d_{M_i} is the death rate of M_i macrophages. The second term on the right-hand side of Eq. (15.2) accounts for the bursting of M_i under bacterial load. The probability for macrophage to burst increase to 50% when the number of internal bacteria reaches N, that is, the burst rate is $\lambda_2/2$ when $B_i = NM_i$. Note that we have assumed here that the transition from non-bursting state to bursting-state is sharp, as in Fig. 10.2(B) rather than Fig. 10.2(A), and so we used the Hill kinetics rather than the Michaelis-Menten law.

We next write a differential equation for the extracellular bacteria:

$$\frac{dB_e}{dt} = N\lambda_2 M_i \frac{B_i^2}{B_i^2 + (NM_i)^2} - \lambda_1 M \frac{B_e}{K+B_e}. \tag{15.3}$$

The first term on the right-hand side accounts for the number of bacteria released at burst of infected macrophages, and the second term represents the loss of B_e due to ingestion by macrophages.

The equation for intracellular bacteria B_i is

$$\frac{dB_i}{dt} = \gamma B_i + \lambda_1 M \frac{B_e}{K+B_e} - N\lambda_2 M_i \frac{B_i^2}{B_i^2 + (NM_i)^2}. \tag{15.4}$$

Here γ is the growth rate of the bacteria within macrophages, and the last two terms in Eq. (15.4) have already been explained above.

We note that the dimension of the parameter λ_1 in Eq. (15.1) is 1/time, but its dimension in Eq. (15.4) is bacteria/(macrophage×time). The same remark applies to the parameter λ_2.

A natural question arises: How can we explain the fact that most infections with Mtb do not lead to chronic active TB? The answer is that the adaptive immune system (located in the lymph nodes) receives stress signals from the infected macrophages M_i, and then inflammatory macrophages (in contrast to non-inflammatory alveolar macrophages) and T cells migrate into the lung and kill bacteria; for simplicity we shall consider only the T cells. Their number, per cm^3, satisfies the equation

$$\frac{dT}{dt} = k_{M_i} M_i - d_T T, \tag{15.5}$$

where d_T is the death rate, and k_{M_i} is the rate by which T cells are activated by the (stress signals sent by the) M_i. In order to take into account the killing of bacteria by T cells we have to replace Eqs. (15.3)–(15.4) by the following equations:

$$\frac{dB_e}{dt} = N\lambda_2 M_i \frac{B_i^2}{B_i^2 + (NM_i)^2} - \lambda_1 M \frac{B_e}{K + B_e} - \delta_1 T B_e, \tag{15.6}$$

$$\frac{dB_i}{dt} = \gamma B_i + \lambda_1 M \frac{B_e}{K + B_e} - N\lambda_2 M_i \frac{B_i^2}{B_i^2 + (NM_i)^2} - \delta_2 T B_i. \tag{15.7}$$

For simplicity we take $\delta_1 = \delta_2 = \delta$. The parameters k_{M_i} and δ determine whether the infection with Mtb will develop into active TB.

The question of susceptibility to TB can be framed as follows: how many ingested bacteria it takes in order to cause an initial infection to develop into a chronic TB? We first address this question with a simple model and later on address it with numerical computations for the full model. The simple model involves only extracellular bacteria B and uninfected macrophages M:

$$\frac{dM}{dt} = M_0 - \mu_1 \frac{MB}{B + K} - \alpha M, \tag{15.8}$$

$$\frac{dB}{dt} = \lambda B - \mu_2 \frac{MB}{B + K}. \tag{15.9}$$

Here M_0 is a baseline supply of new macrophages, α is the natural death rate of macrophages, μ_1 is the rate by which macrophages ingest bacteria, a process that depletes the bacteria at rate μ_2, and λ is a constant. The ingestion process (**endocytosis**) is modeled by the Michaelis-Menten formula. In steady state of healthy individuals, $M_0 - \alpha M = 0$.

The model (15.8)–(15.9) is very simple since, as we know from the more detailed model (15.1)–(15.2), (15.5)–(15.7), that λ is a function of B_i, M_i and T. Nevertheless, already the simple model (15.8)–(15.9) sheds some light on the consideration of susceptibility to TB, as we shall see from the following problems.

Problem 15.1. We may view the system (15.8)–(15.9) as a model of an infectious disease with DFE (disease-free equilibrium)

$$(M, B) = (\frac{M_0}{\alpha}, 0).$$

Setting

$$b = \frac{\mu_2 M_0}{\alpha} - \lambda K,$$

show that the DFE is stable if $b > 0$ and unstable if $b < 0$.

We next study the behavior of solutions of Eqs. (15.8) and (15.9) when the initial values are not necessarily near the DFE.

Problem 15.2. Show that if $M(0) \leq \frac{M_0}{\alpha}$ then $M(t) \leq \frac{M_0}{\alpha}$ for all $t > 0$, and deduce that

$$\frac{dB}{dt} \geq B\frac{\lambda B - b}{B + K}.$$

Problem 15.3. Deduce from Problem 15.1 that if initially $M(0) \leq \frac{M_0}{\alpha}$ and $B(0) > b/\lambda$ ($B(0) > 0$ if $b < 0$) then $B(t) \to \infty$ as $t \to \infty$, which means that infection with more than b/λ bacteria will develop into active TB. [Hint: $\frac{dB}{dt}(0) > 0$, hence $B(t) > B(0)$ for small t. Show that $B(t) > B(0)$ for all $t > 0$, $B(t)$ is monotonically increasing, and

$$\frac{dB(t)}{dt} > \frac{B(0)}{B(0) + K}(\lambda B(0) - b)$$

for all $t > 0$.]

If $b > 0$ then the DFE $(\frac{M_0}{\alpha}, 0)$ is stable. Hence if $M(0)$ is near M_0/α and $B(0)$ is sufficiently small then $B(t) \to 0$ as $t \to \infty$. The following theorem shows that this result remains true also whenever $M(0) > M_0/\alpha$.

Theorem 15.1. *If* $b > 0$, $M(0) > \frac{M_0}{\alpha}$ *and* $B(0) < \varepsilon$ *where* ε *is sufficiently small, then* $B(t) \to 0$ *as* $t \to \infty$.

Thus infection with a small number of *Mtb* will not develop into TB, whenever the DFE is stable and the immune system is strong enough in the sense that $M(0) > M_0/\alpha$.

Proof. We introduce the number

$$\beta = \alpha + \frac{\mu_1 \varepsilon}{\varepsilon + K}.$$

Since $M(0) > \frac{M_0}{\alpha}$ and $b > 0$, if ε is sufficiently small then

$$M(0) > \frac{M_0}{\beta} \quad \text{and} \quad \lambda < \frac{\mu_2 M_0}{\beta(\varepsilon + K)}. \tag{15.10}$$

We claim, that if $B(0) < \varepsilon$ then

$$B(t) < \varepsilon \text{ for all } t > 0. \tag{15.11}$$

We prove this assertion by contradiction. Suppose (15.11) is not true. Then there is a first time t_0 such that

$$B(t) < \varepsilon \text{ if } t < t_0, \text{ and } B(t_0) = \varepsilon.$$

It follow that

$$\frac{dB}{dt}(t_0) \geq 0 \tag{15.12}$$

and

$$\frac{B(t)}{B(t)+K} < \frac{\varepsilon}{\varepsilon+K} \quad \text{if } t < t_0. \tag{15.13}$$

Hence, by Eq. (15.8),

$$\frac{dM}{dt} > M_0 - \mu_1 M \frac{\varepsilon}{\varepsilon+K} - \alpha M = M_0 - \beta M. \tag{15.14}$$

Rewriting this inequality in the form

$$\frac{d}{dt}(Me^{\beta t}) > M_0 e^{\beta t} \tag{15.15}$$

we obtain, by integration,

$$M(t) > M(0)e^{-\beta t} + \frac{M_0}{\beta} - \frac{M_0}{\beta}e^{-\beta t} > \frac{M_0}{\beta} \quad \text{for } 0 < t \le t_0, \tag{15.16}$$

where we used the inequality $M(0) > \frac{M_0}{\beta}$. We now use Eq. (15.9) to deduce that

$$\frac{dB}{dt}(t_0) = (\lambda - \frac{\mu_2 M}{B+K})B|_{t=t_0} < (\lambda - \frac{\mu_2}{\varepsilon+K}\frac{M_0}{\beta})\varepsilon < 0$$

by (15.10), which is a contradiction to (15.12), thus proving the assertion (15.11).

We can now repeat the previous arguments and establish the inequalities (15.13)–(15.16) for any $t_0 > 0$. Hence

$$M(t) > \frac{M_0}{\beta} \quad \text{for all } t > 0.$$

But

$$\frac{1}{\beta} = \frac{1}{\alpha + \mu_1\varepsilon/(\varepsilon+K)} > \frac{1}{\alpha} - C\varepsilon$$

for some constant C. Hence

$$M(t) > \frac{M_0}{\alpha} - C\varepsilon \quad \text{for all } t > 0.$$

On the other hand we deduce from Eq. (15.8) that

$$\frac{dM}{dt} \le M_0 - \alpha M$$

and then, by Theorem 2.3,

$$M(t) < \frac{M_0}{\alpha} + \varepsilon$$

for any $\varepsilon > 0$ if t is sufficiently large, say $t > T_\varepsilon$. Hence

$$-C\varepsilon < M(t) - \frac{M_0}{\alpha} < \varepsilon$$

if $t = t_0$ for any time $t_0 > T_\varepsilon$, while $B(t_0) < \varepsilon$. It follows that $(M(t_0), B(t_0))$ lies in a small circle about the DFE $(\frac{M_0}{\alpha}, 0)$. Since $b > 0$, the DFE is asymptotically stable (by Problem 15.1). If follows that $B(t) \to 0$ as $t \to \infty$, and the proof of Theorem 15.1 is complete.

In summary, the model (15.8)–(15.9) makes the following predictions: (i) If the immune system is weak (i.e., $M(0) < M_0/\alpha$) and the DFE is unstable (i.e., $b < 0$) then any small infection with Mtb could develop into TB. (ii) If the DFE is stable (i.e., $b > 0$) but the immune system is weak, then any infection at level above b/λ will develop into TB. (iii) If the DFE is stable and the immune system is strong (i.e., $M(0) > M_0/\alpha$), then any small infection by Mtb will not develop into active TB.

Problem 15.4. Prove that under the assumptions of Theorem 15.1, $M(t) \to \frac{M_0}{\alpha}$ as $t \to \infty$.

The mathematical model on tuberculosis in this chapter is a simplification of the model from article [10].

15.1 Numerical Simulations

In the following problems the parameters for Eqs. (15.1)–(15.7) are given as follows: $\lambda_1 = 2$/day, $\lambda_2 = 0.05$/day, $d_M = 8 \times 10^{-4}$/day, $M_0 = 1.5 \times 10^6$ cell/cm^3, so that $\mu_M = d_M M_0 = 1.2 \times 10^4$ cell/day, $d_{M_i} = 5 \times 10^{-2}$/day, $K = 10^7 B_e$/cm^3, $\gamma = 0.8$ day. At the beginning of infection with Mtb we have: $M = M_0, M_i = 1, B_i = 25$ and B_e is the number of inhaled bacteria per cm^3. We also take in (15.5) $d_T = 0.3$ /day, $k_{M_i} = 2.5$/day and $T(0) = 0$.

Problem 15.5. Simulate the model (15.1)–(15.4) with $B_e(0) = 100$ for $0 < t < 30$ days. Sample codes are shown in Algorithms 15.1 and 15.2.

Problem 15.6. Use the model (15.1), (15.2), (15.5)–(15.7) with $\delta = 10^{-7}$/day and $B_e(0) = 100, 200, 500, 1000$ to compute $B_e(30)$ and $B_i(30)$.

Problem 15.7. Repeat the calculations of Problem 15.6 with $B_e(0) = 100$. Change δ to 10^{-6} and 10^{-5} to see the effect.

Algorithm 15.1. Main file for simulating Problem 15.5 (main_TB.m)

```
% This code is to simulate Problem 15.5

global lambda_1 lambda_2 dM M0 mu_M dMi K gamma dT k_Mi N

%% parameters
lambda_1 = 2;            % /day
lambda_2 = 0.5;          % /day
dM       = 8 * 10^(-4);  % /day

M0    = 1.5 * 10^6;   % cell/cm^3
mu_M  = dM * M0;      % cell/day
dMi   = 5 * 10^(-2);  % /day
K     = 10^7;         % /cm^3
gamma = 0.8;          % day

dT   = 0.3;           % /day
k_Mi = 2.5;           % /day
N    = 50;

%% initial conditions
M  = M0;      % alveolar macrophages per cm^3
Mi = 1;       % infected alveolar macrophages per cm^3
Be = 100;     % extracellular bacteria per cm^3
Bi = 25;      % intracellular bacteria per cm^3

z_ini = [M; Mi; Be; Bi];
tspan = [0,30];

%% ODE solver
[t,z] = ode45('fun_TB',tspan,z_ini);

%% Plot
tvec = {'M','Mi','Be','Bi'};

for i = 1 : 4
    subplot(2,2,i)
    plot(t,z(:,i))
    xlabel('t'),ylabel(tvec(i))
end
```

Algorithm 15.2. fun_TB.m

```
function dz = fun_TB(t,z)
global lambda_1 lambda_2 dM M0 mu_M dMi K gamma dT k_Mi N

dz = zeros(4,1);

M  = z(1);
Mi = z(2);
Be = z(3);
Bi = z(4);

dz(1) = mu_M - lambda_1*M*Be/(K+Be) - dM*M;
dz(2) = lambda_1*M*Be/(K+Be) - lambda_2*Mi*Bi^2/(Bi^2+(N*Mi)^2) ...
            - dMi*Mi;
dz(3) = N*lambda_2*Mi*Bi^2/(Bi^2+(N*Mi)^2) - lambda_1*M*Be/(K+Be);
dz(4) = gamma*Bi + lambda_1*M*Be/(K+Be) ...
            - N*lambda_2*Mi*Bi^2/(Bi^2+(N*Mi)^2);
```

Solutions

Problems of Chapter 2

2.1 (i) $x = ce^{-t} + \frac{3}{2}e^{t}$; (ii) $x = ce^{-t^2} + \frac{1}{2}$; (iii) $x = \frac{t^2}{\alpha+2} + \frac{C}{t^\alpha}$.

2.2 (i) $x = 3e^{-t^2/2} - 1$; (ii) $-\frac{2}{9}e^{3t} - \frac{t}{3} - \frac{7}{9}$.

2.3 (i) $x^2 = t^2 + 8$; (ii) $\frac{1}{2}\ln(1+x^2) = \ln t + \frac{1}{2}\ln 5$.

2.4 $3\ln|x - 2t| + \ln|x + 2t| = 4\ln 3$.

2.5 $x^2 t + \ln|x| = 3$.

2.6 $x^3 + t^2 x = 8$ (using integrating factor $\frac{1}{x}$).

2.7 $x = a$ is stable.

2.8 The solution is

$$\frac{1}{2-a}\left[\ln|a - x| - \ln|2 - x|\right] = t + C.$$

Problems of Chapter 3

3.1 Special solution: $-t^2 + 2t + 4$.

3.2 $x(t) = c_1 e^t + c_2 e^{3t} + \frac{1}{8}e^{-t}$, $c_1 + c_2 = 0$, $c_1 + 3c_2 = \frac{1}{8}$.

3.3 $c_1 e^{5t} \begin{pmatrix} 1 \\ 1 \end{pmatrix} + c_2 e^{-4t} \begin{pmatrix} 7 \\ -2 \end{pmatrix}$.

© Springer International Publishing Switzerland 2016
C.-S. Chou, A. Friedman, *Introduction to Mathematical Biology*,
Springer Undergraduate Texts in Mathematics and Technology,
DOI 10.1007/978-3-319-29638-8

3.4 Take real and imaginary parts of $e^{(1+2i)t} \begin{pmatrix} 1 \\ -i \end{pmatrix}$: real part is $\begin{pmatrix} e^t \cos 2t \\ e^t \sin 2t \end{pmatrix}$ and imaginary part is $\begin{pmatrix} e^t \sin 2t \\ -e^t \cos 2t \end{pmatrix}$.

3.5 $(e^t - e^{-t}, -2e^t + e^{-t})$, $(e^t + e^{-t}, -2e^t - e^{-t})$.

3.6 $(2e^{-t}\cos 3t, e^{-t}(\cos 3t - \sin 3t))$, $(2e^{-t}\sin 3t, e^{-t}(\cos 3t + \sin 3t))$.

3.8 Special solution $(\frac{1}{2}e^{-t}, 2e^{-t})$.

Problems of Chapter 4

4.1 Eigenvalues $2, -1$ at $(1,1)$; $1, -2$ at $(-1,1)$.

4.2 Eigenvalues $0, 1$ at $(0, -1)$; $1, -4$ at $(-2, 1)$.

4.3 Equilibrium points
$(x,y) = (1 \pm \sqrt{2}, 1), (-1 \pm \sqrt{2}, -1)$; only $(-1 - \sqrt{2}, -1)$ is stable.

4.4 $(0,0), (0, \frac{1}{4}), (1,0)$ – all unstable.

Problems of Chapter 5

5.1 $(0,0)$ and $(A,0)$ are unstable; $(0,1)$ is stable if $a < b$, unstable if $a > b$. $\bar{x} = (a-b)/(b+\frac{a}{A}), \bar{y} = 1 + \bar{x}$ is stable if $a > b$, i.e., if the growth rate of the prey is larger than the rate b by which the prey is killed by the predator.

5.2 The eigenvalues are $-\alpha r$ and 0.

Problems of Chapter 6

6.1 $(0,0)$ is unstable; $(k_1,0)$ is stable if $k_1 > \frac{r_2}{b_2}$; $(0,k_2)$ is stable if $k_2 > \frac{r_1}{b_1}$.

Problems of Chapter 7

7.1 Steady point

$$\bar{x} = \frac{r_1}{k_1}, \quad \bar{z} = 1 - \frac{r_1 \gamma_2}{r_2 \gamma_1}, \quad \bar{y} = \frac{1}{\beta}[\mu - r_1(1 - \frac{\bar{x}}{A}) - \beta_2 \bar{z}].$$

Compute the characteristic polynomial without substituting the specific values of $\bar{x}, \bar{y}, \bar{z}$.

7.2 Steady point

$$\bar{x} = k_1\left(1 + \frac{\beta_1}{r_1}\bar{z}\right), \quad \bar{y} = k_2\left(1 + \frac{\beta_2}{r_2}\bar{z}\right), \quad \bar{z} = \frac{B}{\alpha}(\alpha - r_1\bar{x} - r_2\bar{y}).$$

Compute the characteristic polynomial without substituting the specific values of $\bar{x}, \bar{z}, \bar{y}$.

7.3 $(A,0,0)$ and $\frac{c_1}{d_1}, \frac{a}{b}(1 - \frac{c_1}{Ad_1})$ are stable, the other two steady points are unstable.

Problems of Chapter 8

8.2 $(\frac{1}{\mu}, 0, \beta - \frac{1}{\mu})$ is stable; $(1, \beta - 1, 0)$ and $(\beta, 0, 0)$ are unstable.

8.3 Unstable.

Problems of Chapter 9

9.2 Substitute $R = \frac{v}{\gamma + \mu}I$, $E = \frac{v + \mu}{k}I$, $S = \frac{(k+\mu)E}{\beta I} = \frac{k+\mu}{\beta}\frac{v+\mu}{k}$ into the first equation in (9.5).

9.6 Use the Routh-Hurwitz criterion.

Problems of Chapter 10

10.3 Dropping the brackets "[ ]" we have:

$$\frac{dE}{dt} = (k_{-1} + k_2)C_1 - k_1SE,$$

$$\frac{dC_2}{dt} = k_3SC_1 - (k_{-3} + k_4)C_2,$$

$$\frac{dC_1}{dt} = [k_1SE - (k_{-1} + k_2)C_1] - [k_3SC_1 - (k_{-3} + k_4)C_2],$$

where the last expression in brackets is dC_1/dt. Since

$$\frac{d}{dt}(E + C_1 + C_2) = 0, \quad E = e_0 - C_1 - C_2, \quad e_0 = constant.$$

$\frac{dC_2}{dt} = 0$ gives $C_2 = \frac{SC_1}{K_2}$, $K_2 = \frac{k_{-3} + k_4}{k_3}$; $\frac{dC_1}{dt} = 0$ gives $k_1S(e_0 - C_1 - C_2) = (k_{-1} + k_2)C_1$, or

$$Se_0 - SC_1 - SC_2 = K_1C_1, \quad K_1 = \frac{k_{-1} + k_2}{k_1}.$$

Hence

$$Se_0 = SC_1 + S\frac{SC_1}{K_2} + K_1C_1 = C_1\left[S + \frac{S^2}{K_2} + K_1\right],$$

so that

$$C_1 = \frac{e_0K_2S}{K_1K_2 + K_2S + S^2} \quad \text{and } C_2 = \frac{e_0S^2}{K_1K_2 + K_2S + S^2}.$$

Now use

$$\frac{dP}{dt} = k_2C_1 + k_4C_2.$$

Problems of Chapter 12

12.2 Stable if

$$\frac{k_1K_1}{(K_1 + \bar{L})^2} < \frac{\lambda}{(H_0/r_2) - \delta}.$$

Problems of Chapter 13

13.1

$$\bar{M}_1 = \frac{k_1}{\mu}, \quad \bar{T} = \frac{k_T}{\mu_T}\frac{k_1}{\mu K_2}, \quad \bar{C} = C_0\left(1 - \frac{k_T\mu_Ck_1}{\lambda_C\mu_T\mu K_2}\right).$$

Problems of Chapter 14

14.1

$$\bar{h} = \frac{\lambda_2}{\mu_2}\bar{c}, \quad \bar{e} = \frac{\lambda_3\bar{h}}{\mu_3\bar{c} + \mu_4},$$

$\bar{c}$ satisfies: $\frac{1}{K}\bar{c}^2 - \alpha\bar{c} + \beta = 0$, where α, β are positive constants independent of K.

14.3 The equilibrium point is not asymptotically stable since the coefficient of λ in the characteristic polynomial is zero.

References

1. Coddington, E.A.: An Introduction to Ordinary Differential Equations. Dover, New York (1989)
2. Butcher, J.C.: Numerical Methods for Ordinary Differential Equation. Wiley, New York (2008)

3. Strang, G.: Linear Algebra and Its Applications, 4th edn. Brooks/Cole, Belmont (2005)
4. Gantmacher, F.R.: The Theory of Matrices, vol. 2. Chelsea, New York (1959)
5. van den Driessche, P., Watmough, J.: Further notes on the basic reproduction number. In: Mathematical Epidemiology, pp. 159–178. Springer, Berlin/Heidelberg (2008)
6. Hale, J.K., Kocak, H.: Dynamics and Bifurcations. Springer, New York (1991)
7. Friedman, A., Hao, W., Hu, B.: A free boundary problem for steady small plaques in the artery and their stability. J. Diff. Eqs. **259**, 1227–1255 (2015)
8. Louzoun, Y., Xue, C., Lesinski, G.B., Friedman, A.: A mathematical model for pancreatic cancer growth and treatment. J. Theor. Biol. **351**, 74–82 (2014)
9. Friedman, A., Tian, J.P., Fulci, G., Chiocca, E.A., Wang, J.: Glioma virotherapy: the effects of innate immune suppression and increased viral replication capacity. Cancer Res. **66**, 2314–2319 (2006)
10. Day, J., Friedman, A., Schlesinger, L.S.: Modeling the immune rheostat of macrophages in the lung in response to infection. Proc. Natl. Acad. Sci. USA **106**, 11246–11251 (2009)

Index

© Springer International Publishing Switzerland 2016   171
C.-S. Chou, A. Friedman, *Introduction to Mathematical Biology*,
Springer Undergraduate Texts in Mathematics and Technology,
DOI 10.1007/978-3-319-29638-8

Printed in the United States
By Bookmasters